Using Global Positioning
Systems in the Outdoors

GPS
MADE
EASY

Fifth Edition

LAWRENCE LETHAM
ALEX LETHAM

W9-BMO-728

THE MOUNTAINEERS BOOKS

THE MOUNTAINEERS BOOKS
is the nonprofit publishing arm of The Mountaineers Club, an organization founded in 1906 and dedicated to the exploration, preservation, and enjoyment of outdoor and wilderness areas.

1001 SW Klickitat Way, Suite 201, Seattle, WA 98134

Published simultaneously in Canada by Rocky Mountain Books, P.O. Box 468, Custer, WA, 98240
Distributed in Europe by Cordee, 3A De Montfort St., Leicester LE1 7HD, Great Britain

Manufactured in Canada

A catalog record for this book is available at the Library of Congress.

♻ This book has been produced on 100% post-consumer recycled paper, processed chlorine free and printed with vegetable-based dyes.

Contents

Acknowledgements

Thank you to Natural Resources Canada for permission to use the Canadian maps on pages 75, 91 and 100. These maps are based on information taken from the National Topographic System map sheet numbers 83 E/3 Mount Robson © 1980 and 83 C/3 Columbia Icefields, edition 02 © 1984 Her Majesty the Queen in Right of Canada with permission of Natural Resources Canada.

The maps shown on pages 13, 26, 28, 29, 70, 89, 103, 105, 107, 109, 111, 142 and 144 were produced by the U.S. Geological Survey. The marine charts on pages 118, 120 and 122 are from the U.S. National Oceanic & Atmospheric Administration.

Thank you to Martin Fox of MapTech for loaning me the Terrain Navigator Pro program. The screen shots of Chapter 9 and part of Chapter 10 are from the MapTech program.

The screen shots from the end portion of Chapter 10 and Chapter 11, and throughout many portions of the book, are of Garmin GPS receivers.

The satellite image in Chapter 7 is reproduced with permission from TerraMetrics, Inc., www.TruEarth.com.

Thank you to Brad Dameron of Garmin, Angela Linsey-Jackson and Lonnie Arima of Magellan, Steve Wegrzyn of Lowrance, Russ Graham of Navman and Cindy Estridge of Pharos for loaning me receivers, software, pictures and accessories produced by their respective companies. Pictures of the products they loaned me and screen shots are found throughout the book.

Trademark List

GPSmap 60CSx, Nuvi, 18 USB, iQue, Rhino, Astro, USB 35 TracPak and MetroGuide USA are trademarks of Garmin International. eXplorist 600, SporTrack and Meridian are trademarks of Magellan, Inc. TOPO! is a trademark of National Geographic. Navman, iCN 720, Navman E-series and Navman i-Series are trademarks of Navman. Terrain Navigator, Terrain Navigator Pro, Pocket Navigator, Chart Navigator and the CAPN are trademarks of Maptech, Inc. Google Earth is a trademark of Google, Inc. TruEarth is a trademark of TerraMetrics, Inc. SanDisk is a trademark of SanDisk Corporation. iPAQ is a trademark of Hewlett-Packard. Bluetooth is a trademark of Bluetooth SIG. TomTom Navigator is a trademark of TomTom NV. Telenav GPS Navigator is a trademark of Telenav Inc. AirClic MP is a registered trademark of Sprint Nextel. Pharos GPS Phone 600 is a trademark of Pharos Science and Applications, Inc. iFinder and Lowrance are trademarks of Navico, Inc. GPS Utility is a trademark of GPS Utility Limited.

1 Introduction to the Global Positioning System (GPS)

The Global Positioning System (GPS) is a satellite system used for navigation. It enables anyone on the planet who owns a GPS receiver to know where they are, 24 hours a day and in any kind of weather.

The GPS is a group of 24 satellites that circle the earth and beam radio signals from their positions back to the earth's surface. A GPS receiver is an electronic device that detects these radio signals and calculates the receiver's position on the earth. GPS affects everyday life. GPS receivers are in cell phones, so when you make an emergency call, your position may automatically be sent to rescuers. Many cars have built-in navigation systems and security systems (e.g., OnStar) that use GPS. GPS receivers can be mounted on bicycles, ATVs, quads, trucks, even an individual's wrist. The ways GPS is used to make life easier will only increase.

The GPS system has radically changed navigation for the outdoor adventurer. A recent improvement called Wide Area Augmentation System (WAAS) has increased the accuracy of GPS receivers to 3 m (9.8 ft.). Improved GPS receivers have downloadable electronic maps that have all the detail of the best topographic maps available. For the urban explorer, electronic city maps give street-by-street directions to your destination. Navigation has never been easier in any environment.

This book will help you understand both the satellite system and GPS receivers, so you will know how to choose the receiver you need and how to use it properly.

Before describing how GPS works, it is interesting to see how hard navigation was in the 15th century. The comparison between then and now is dramatic.

GPS technology is based on a group of satellites that beam radio signals back to earth.

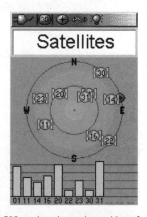

A GPS receiver shows the position of each satellite and signal strength bars.

Early Navigation

The times leading up to Columbus' voyage to the New World in 1492 was an era of navigational advances. The Greeks and Arabs of antiquity knew how to measure latitude from the stars. Sailors would note their latitude when leaving a port and return to port by simply sailing the opposite direction at the same latitude.

Latitude was relatively easy to measure because one simply had to measure the sun's arc in the sky to know one's distance from the equator. The quadrant was the tool of choice for determining latitude, and Columbus used one when he sailed. But there was no easy way to determine longitude, because it required celestial knowledge and math skills that most sailors did not have. Amerigo Vespucci was an amazing exception in his day. His extensive knowledge of the stars and his mathematical abilities enabled him to determine longitude and to make an amazing discovery.

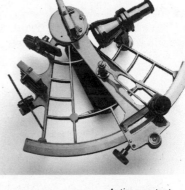

Antique sextant

In 1499, at the age of 48, Vespucci sailed from the Old World to the lands that Columbus had discovered and claimed were India. Vespucci carried with him an almanac made in Ferrara, Italy, that showed the positions of various stars at exact times. The almanac showed that on August 23, 1499, the moon would cross Mars at exactly midnight in Ferrara. Vespucci had himself put ashore on what is today the Brazilian coast. He measured the stars to determine his exact local time and waited for the conjunction between the moon and Mars. When the moon finally crossed Mars in Brazil, it was 6.5 hours after it was observed in Ferrara.

Using the earth's circumference as calculated by Claudius Ptolemy in 140 CE (17,895 miles as opposed to 24,900 miles), Vespucci determined that he was about 4800 miles west of Ferrara. Vespucci had calculated his longitude. He knew where he was on the globe. He also knew that Columbus had discovered not a new route to the Indies, but an entirely new world.

John Harrison made calculating longitude a lot easier. Harrison produced the first accurate chronometers in 1761. Ships carried chronometers set to the time of their home ports, usually Greenwich, England. The difference between local time (measured by the stars) and home-port time (shown by the chronometer) was the distance, as per the earth's circumference, from the home port. Captain James Cook used a Harrison-type chronometer to measure longitude.

Unfortunately for both Vespucci and Captain Cook, if the sky was cloudy it was impossible to determine position, because celestial observation was still an indispensable part of navigation.

Today, finding latitude and longitude is as simple as turning on a GPS receiver. GPS does not remove the need to do math or to know the positions of heavenly objects, but the system does all the hard work. Knowing your exact position is as easy as looking at the GPS receiver display. Even better, GPS receivers work when it is cloudy, raining, snowing, or even in the smoke-filled air of a forest fire.

Even though GPS users do not need to understand the system to use it, some knowledge of the system's intricacies is helpful.

How GPS Works

The U.S. Air Force conceived the GPS in 1960 as a method to make intercontinental ballistic missiles more accurate. By 1974, the other branches of the U.S. military had joined the project and renamed it Navstar, though the new name never caught on. The system cost $10 billion to develop and became fully operational in April 1995.

The number of satellites in orbit fluctuates between 24 and 29. Eighteen satellites are the necessary minimum to provide worldwide coverage, but additional satellites were launched as spares and for upgrading the system.

The GPS is accurate to about 15 m (49.2 ft.). In 1998, the Wide Area Augmentation System (WAAS) improved accuracy to better than 3 m (9.8 ft.). WAAS was added to increase accuracy for commercial airlines. WAAS

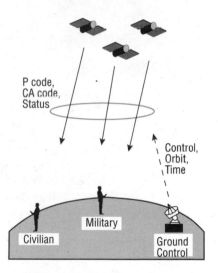

The three parts of the GPS: ground, space and user.

and other accuracy-improving systems are explained in Chapter 2.

The actual workings of the system are amazing. The system is divided into three segments: space, ground control and users. The space segment includes the satellites that orbit 20,200 km (12,552 mi.) above the earth. The satellites beam radio signals toward earth at 1227.6 MHz (L2 frequency) and 1575.42 MHz (L1 frequency). The radio signals broadcast an almanac and an electronic code. The almanac is the position of each satellite around the world. The electronic code is a long string of ones and zeros. Each satellite broadcasts a unique code which permits a GPS receiver to measure its distance from that particular satellite. The satellites actually broadcast two codes: a precision code (P code, accessible only to the military) and a coarse acquisition code (CA code, used by civilians).

Each satellite has a highly accurate atomic clock. The ground segment keeps the atomic clocks adjusted so that each satellite maintains precisely the same time. The codes are sent from the satellites at the same time. As will be seen, keeping accurate time is very important to the GPS.

The ground control segment has stations in Hawaii, Colorado, Ascension Island in the South Atlantic, Diego Garcia in the Indian Ocean, and Kwajalein in the Pacific. These ground stations track the satellites, monitor their health, and make any necessary adjustments to the atomic clocks. The U.S. Department of Defense controls the system.

Modern GPS receivers are small, lightweight, and run on two AA batteries.

GPS receivers make up the user segment. It is the GPS receiver, whether located in an airplane, a truck or in a hiker's hand, that detects the radio signals from the satellites and calculates the receiver's position. The number of users does not affect the radio signals, so every person in the world could have a receiver and the system would still function properly.

The way a GPS receiver uses the radio signals to calculate its position is a work of ingenuity. When a receiver is first turned on, it receives the almanac, so the receiver knows the position of all of the satellites even if a satellite is on the other side of the world. The receiver then receives the exact time. The crystal oscillator in a GPS receiver is not accurate enough to calculate position, so the receiver receives the time kept by the atomic clocks.

The receiver monitors four of the satellites that are currently overhead. At least four satellites should be detectable from any spot on the globe at all times. One satellite provides the receiver with the system time from the atomic clocks. The other three satellites transmit their codes from orbit. The receivers transmit their code at the same time. The receiver detects when it receives the code from each satellite. The time difference between when a code was sent and when it was received is the time the code took to travel from the satellite to the receiver. The speed of the satellite signals is known, so the receiver calculates its distance from three celestial objects (the satellites) to determine its three-dimensional position (e.g., latitude, longitude, altitude) on the earth.

GPS is just like navigating by the stars. If you know the position of the stars, you can calculate your position on the earth. The satellites are like stars, except they radiate a radio signal detectable by a GPS receiver instead of light detectable by the human eye.

Another way to understand how GPS works is to imagine yourself floating in a room of zero gravity. All the walls look the same, so you use a tape measure to determine the distance to three walls. Because the positions of the walls are known, measuring your distance from three walls tells you your position in the room. GPS is much the same. The receiver

Some small GPS receivers use lithium-ion rechargeable batteries.

knows the position of three satellites, even though they are constantly moving, and measures the distance to each satellite to calculate its own position. The system is very complex, yet it is simple to use.

Additional Terms and Details

You do not need to understand this section to use a GPS receiver, but being familiar with GPS terms and concepts helps when reading other GPS material.

The P code and the CA code are composed of binary numbers (strings of 0s and 1s). The P code is so long that it takes seven days to repeat itself. The CA code repeats itself every millisecond. The P code also transmits more bits per second (chipping rate) than the CA code, which means that

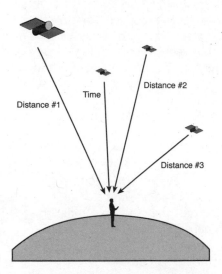

One satellite is required for synchronizing time, and three more for a 3D position fix.

it changes more frequently than the CA code. If the P and CA codes were compared to the markings on a tape measure, the P code would have millimeter (1/32 inch) marks, whereas the CA code would have only centimeter (1/3 inch) marks. The finer marks of the P code allow it to be more accurate, but the bad news is that only military receivers understand the P code. Civilian receivers access only the CA code. But the news is not really all that bad. While military receivers are accurate under most conditions to 1 m (3.28 ft.), a civilian receiver with WAAS is accurate to 3 m (9.8 ft.), which for most adventurers is fabulous accuracy.

The ionosphere affects GPS accuracy. The ionosphere is the layer of atmosphere 150 to 900 km (93 to 559 mi.) above the earth. Radio waves slow down in the ionosphere. As you recall, a GPS receiver uses the speed of the radio wave to calculate the distance to a satellite. If radio waves slow down in the ionosphere, the receiver needs to know how much the signal slowed down, so it can adjust its position calculation.

There are three ways to know how much the ionosphere affects a signal. The first method is accessible only to military receivers. The P code transmits on both the L1 and the L2 frequencies. Radio waves of different frequencies slow down different amounts as they pass through the ionosphere. Military receivers measure the difference in delay between L1 and L2 signals and calculate the amount the signal slowed in the ionosphere.

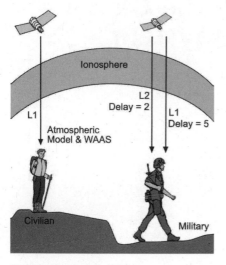

Correction for ionospheric delays.

Military receivers accurately compensate for the ionosphere because they measure the actual delay and can accurately compensate.

The CA code transmits on only the L1 frequency, so civilian receivers use a second method to detect delay – an ionospheric model. The ionosphere is different all over the earth and varies throughout the day, so a receiver cannot simply assume that the ionosphere will slow a radio wave down by a fixed amount. An ionospheric model estimates delay through the ionosphere. Models are never 100% accurate, so civilian receivers cannot fully determine or compensate for ionospheric delay using models.

WAAS is the third method for correcting for ionospheric delay. WAAS uses differential GPS (DGPS) to remove the error introduced by the ionosphere. DGPS is explained in detail in Chapter 2.

As you learn about GPS, you may hear the term Selective Availability (SA). When the GPS was developed, the U.S. government feared the system would be used against it, so the Department of Defense deliberately degraded the accuracy of the CA code. Under SA, the accuracy of a civilian receiver varied randomly between 15 m (49.2 ft.) and 100 m (328 ft.). SA was deactivated May 2, 2000, and civilian receivers became nominally accurate to 15 m (49.2 ft.) without WAAS. Although SA is no longer a factor in GPS accuracy, it could conceivably be reapplied.

2 Accuracy of the Global Positioning System

A GPS receiver excels at three basic tasks:

1. Leading you to a destination you choose from a map.
   The map can be:
   - A paper map
   - An electronic map loaded into the receiver
   - An electronic map displayed on your computer

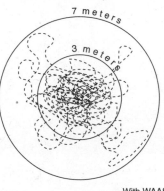

With WAAS

2. Determining your current position, which will be displayed as:
   - A coordinate you can look up on a paper map
   - A location on the electronic map in the receiver
   - A location on the electronic map displayed on your computer

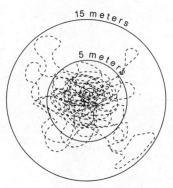

Without WAAS

3. Remembering your current position, so you can return to it later

95% of the time, GPS accuracy is 5 meters without WAAS and 3 meters with WAAS.

How well a receiver performs the above tasks depends entirely on the receiver's accuracy. So, you may ask, "How accurate is my receiver?" The short answer is:

Without WAAS (Wide Area Augmentation System)

- 15 m (49.2 ft.) horizontal accuracy
- 19 m (62.3 ft.) vertical accuracy

With WAAS

- 3 m (9.8 ft.) horizontal accuracy
- 6 m (19.7 ft.) vertical accuracy

If you were to stand in the same location for several hours or days and make a plot of the positions reported by your receiver, the plot would look like the circles pictured opposite. Even though you did not move, the receiver reported different positions because factors like the ionosphere and satellite geometry affected the receiver's accuracy. Accuracy is generally specified as how close the receiver will get you to the center of the circle 95% of the time. A receiver using WAAS, in ideal conditions, is accurate to 3 m (9.8 ft.) 95% of the time and is never less accurate than 7 meters (22.97 ft.).

A geographical representation of accuracy is shown in the figure below. The map is the U.S. Geological Survey map of Mirror Lake, Utah (scale 1:24,000) with grid lines 1000 m (3281 ft.) apart. Three circles, one near Scout Lake, a second southwest of Camp Steiner and the third on the road, are each 30 m (98.4 ft.) in diameter, which means the edge

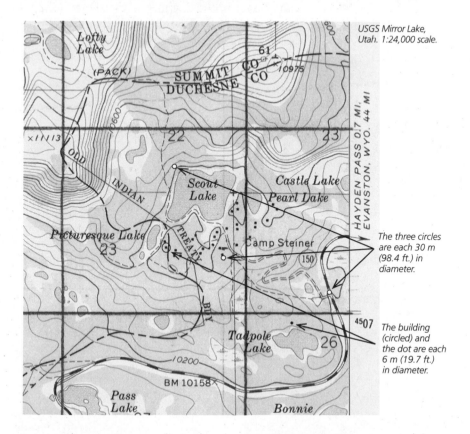

USGS Mirror Lake, Utah. 1:24,000 scale.

The three circles are each 30 m (98.4 ft.) in diameter.

The building (circled) and the dot are each 6 m (19.7 ft.) in diameter.

of the circle is 15 m (49.2 ft.) from the center point. If you wanted to go to one of these points and you did not have a WAAS-enabled receiver, you would get somewhere within the circle. The dot just north of Tadpole Lake and the building (circled) just south-southwest of Scout Lake are approximately 6 m (19.7 ft.) in diameter, or 3 m (9.8 ft.) from the center point. A WAAS-enabled receiver would take you somewhere inside the points.

The map shows that even without WAAS, a GPS receiver makes navigation quite accurate. WAAS only improves accuracy. Understanding WAAS requires an understanding differential GPS.

Differential GPS

Just as the name suggests, Differential GPS (DGPS) uses the difference between two measurements to improve accuracy. DGPS can increase accuracy to 2 cm (0.79 in) when necessary, but as you can imagine, greater accuracy means extra equipment and increased cost.

Real-time corrections are made while you are moving. They could be described as "instantaneous" as opposed to after-the-fact. Post-processing corrections are made in the office after outdoor GPS measurements are complete. Post-processing increases accuracy by analyzing collected data after the fact. DGPS provides real-time corrections that improve GPS receiver accuracy while you are in the field.

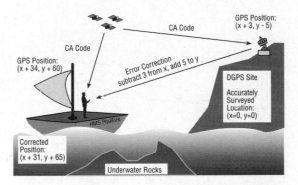

DGPS, available in most U.S. ports, increases civilian receiver accuracy.

This figure of a boat illustrates how real-time DGPS works. The boat captain's chart shows an underwater rock somewhere in the channel. He needs to avoid the rock or it will sink his vessel. The captain has a GPS receiver that can receive real-time differential corrections to improve the accuracy of the positions reported by his receiver.

A nearby DGPS site has a GPS receiver, a computer and a radio transmitter. When the DGPS site was built, the location of the receiver was accurately surveyed to be (x, y), so the DGPS computer accurately knows the receiver's location. The receiver at the DGPS site continuously calculates its position, yet it never moves. When the receiver reports its position as (x+3, y–5), the computer knows that the reported position is inaccurate, and it calculates the amount of error by taking the difference between the reported position and the receiver's known position. In this case, the receiver's positional error is +3 units in the x direction and –5 units in the y direction. The amount of error is reported to the captain's GPS receiver by radio.

The captain's receiver uses the correction information to make its position calculation more accurate. For example, if the captain's receiver reports a position of (x+34, y+60), the information from the DGPS site corrects the position to be (x+31, y+65). The differential correction information makes the captain's receiver more accurate than the nominal accuracy of 15 m (49.2 ft.).

An underlying assumption of DGPS is that the receiver at the DGPS site and the captain's receiver use the same satellites to calculate their positions. Two receivers generally have to be within about 274 km (170 mi.) of each other to use the same satellites, so DGPS has some inherent geographical restrictions. However, WAAS has overcome the distance limitation, so that a receiver can use WAAS correction information over a wide area. WAAS also broadcasts its correction information over the L1 and L2 frequencies, so the captain's receiver does not need an extra radio to receive the WAAS correction information.

What is WAAS?

WAAS stands for Wide Area Augmentation System. Geographically separated DGPS sites use differential GPS techniques to calculate the amount of error in the GPS signals for the satellites it detects. Correction information is transmitted from the various DGPS sites to one of two WAAS satellites stationed over the equator. The WAAS satellites transmit the correction information for all satellite signals to GPS receivers located over an area that is larger than the area covered by a single DGPS site.

When using WAAS, a receiver calculates its position using the satellite signals in the area where it is located and uses information from one of the WAAS satellites to correct its calculated position. WAAS does not help a receiver determine its location, but it provides data to help the receiver refine its position calculation. WAAS testing in September 2002 confirmed accuracy of 1–2 meters horizontal and 2–3 meters vertical throughout most of the continental U.S., much of Alaska and some of Canada. WAAS was developed to increase GPS accuracy for airport landing systems.

FAA documents state that the accuracy using WAAS is 7 m (23 ft.), yet GPS receiver manufacturers claim the accuracy is 3 m (9.8 ft.).

As mentioned above, accuracy is generally specified as accuracy over a percent of time. WAAS makes a receiver accurate to 3 m (9.8 ft.) 95% of the time. The remaining 5% of the time, the receiver is accurate to 7 m (23 ft.). In other words, if you use your receiver to go to the same location 100 times, 95 times you will arrive within 3 m (9.8 ft.) of the destination. The other five times you will arrive within 7 m (23 ft.) of the destination.

The great thing about WAAS is that it broadcasts on the same frequencies used by the GPS satellites (L1 and L2), so no additional equipment besides a WAAS-enabled receiver is required to receive the WAAS corrections. Currently there are 38 reference stations, four uplink stations and two geosynchronous satellites to provide WAAS correctional data. Five reference stations are located in Mexico and four in Canada to provide WAAS coverage in those countries and improve availability in the U.S. WAAS covers altitudes of up to 100,000 feet (18.9 mi., 30.5 km). By 2013, the U.S. government will add a new L5 frequency to provide increased accuracy for aviation. WAAS gets better every year.

WAAS Coverage

WAAS coverage is available inside the circle.

However, if you travel in an area not covered by WAAS, be sure to turn WAAS off. If a receiver picks up WAAS signals in an area that WAAS does not cover, the receiver's accuracy can be adversely affected and result in accuracy worse than 15 m (49.2 ft.).

Other WAAS Systems

Luckily for European travelers, the European Geostationary Navigation Overlay Service (EGNOS) provides corrections for accuracy comparable to WAAS. EGNOS currently supplements the U.S. GPS and will supplement the Russian GLONASS and European Galileo systems when they are completed.

Other WAAS systems under development are MSAS in Japan, GAGAN in India and GRAS in Australia. All three of these systems are targeted toward increasing accuracy for air traffic control.

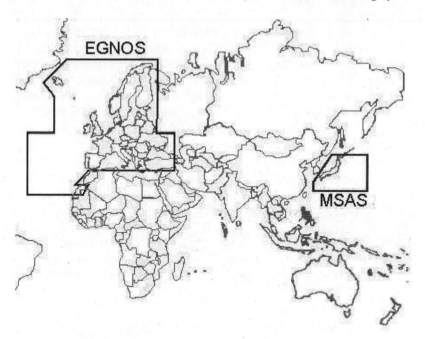

EGNOS coverage for Europe and MSAS coverage for Japan.

Local Area DGPS

Localized DGPS covers small areas, but provides position accuracy that is better than WAAS. Localized DGPS is usually limited to seaports and airports. The systems are designed to guide ships and land aircraft.

Most seaport DGPS correction information is broadcast over the Loran-C frequency. To receive the signals, you need to purchase a Loran-C antenna and a special DGPS receiver box to connect to your GPS receiver. DGPS corrections are usually sent in the RTCM format, so if you want to use DGPS, be sure your receiver accepts the RTCM format.

The U.S. Coast Guard operates more than 60 DGPS sites along U.S. coasts, the Great Lakes, Puerto Rico, Hawaii and parts of Alaska. The Coast Guard DGPS effort was merged into the Nationwide Differential GPS (NDGPS) program, which planned to provide ground-based coverage, as opposed to the satellite-based WAAS coverage, across the continental U.S., Alaska and Hawaii. The funding for NDGPS was cut to zero in fiscal 2006, so the survival of ground-based DGPS is in question.

Even if NDGPS does not survive, most outdoor users will not care, because WAAS is more convenient. Other countries operate ground-based DGPS systems, but those same countries, in Europe for example, are adding their own WAAS corrections systems (e.g., EGNOS). The corrections provided by WAAS should meet most recreational outdoor GPS users' needs.

The U.S. Federal Aviation Administration (FAA) is currently developing a system for aircraft landings at small and medium-sized airports called the Local Area Augmentation System (LAAS). It operates within a 20- to 30-mile radius, broadcasting a correction message via a VHF radio datalink from a ground-based transmitter. LAAS currently provides Category I precision landing and has the goal of providing Category II and III precision landings in the future. LAAS is presently being tested at six U.S. airports and requires higher-end aviation receivers.

One fascinating application of DGPS equipment is tracking the movements of hydrological dams. As a lake fills with water, the DGPS equipment reports how much the dam flexes, which allows the operators to keep the water levels and the fill rate below the failure point of the dam.

More on Accuracy

In areas of optimal satellite coverage and geographic surroundings, the simple answers given above as to a receiver's accuracy are generally true. However, accuracy also depends on:

- ionospheric interference
- satellite geometry
- reflected, or multipath, signals

Ionospheric Interference

Ionospheric interference was described in Chapter 1. Basically, the radio signals from the satellites slow down in the ionosphere. The speed of the signals is important in calculating position. If the receiver can determine the amount the signals slowed in the ionosphere, it can calculate a more accurate position. To the extent that the receiver cannot determine the amount the signal slowed down, the position calculation is inaccurate. Ionospheric delay accounts for 5 to 10 m (16.4 to 32.8 ft.) of error of the total 15 m accuracy.

Military receivers measure ionospheric delay by comparing the delay of different frequencies (L1 and L2). Civilian receivers use mathematical models of the ionosphere to predict delay. WAAS corrects much of the error introduced by the ionosphere.

Satellite Geometry

Satellite geometry, also known as satellite constellation, refers to the satellites' positions in the sky relative to your position. Some geometries enable the receiver to more accurately calculate position. The ideal geometry is one satellite directly overhead and the three other satellites evenly spaced around the horizon.

The amount of error introduced by satellite geometry is called dilution of precision (DOP). DOP has several components: vertical, horizontal, time, position and geometric. A receiver calculates each component of DOP for each combination of four satellites in view. The receiver uses the signals from the four satellites that provide the lowest position dilution of precision (PDOP) number.

Poor geometry may increase position error by hundreds of meters. PDOP values between 1 and 3 provide 15 m (49.2 ft.) accuracy. PDOP values between 4 and 6 may cause inaccuracies from tens of meters to hundreds of meters. Generally, a receiver cannot lock when the PDOP value is greater than 6, thereby resulting in what is known as an outage. An outage may also occur if local terrain blocks the satellites capable of providing the lowest PDOP value. The most you can do when poor satellite geometry results in a position error or an outage is to wait for the constellation to change. Fortunately the satellites are constantly moving, so an outage that is not caused by terrain should last only a few minutes.

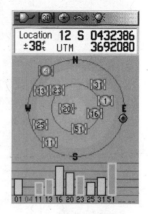

Most receivers do not display DOP values, but they do provide an Estimated Position Error (EPE) that is an indication of the PDOP. EPE shows the amount of error due to the satellite geometry. As you navigate, occasionally look at the EPE to determine if the position error is greater than the ideal values given above.

A receiver's mask angle and antenna sensitivity can help minimize outages. Receivers improve accuracy by ignoring satellites that are close to the horizon. Mask angle refers to the number of degrees a satellite must be

EPE reading from a receiver is + 38 ft.

above the horizon before it is used to calculate a position. Most receivers have a mask angle between 5° and 10°. The larger the mask angle, the more a receiver is affected by outages because the receiver stops using satellites near the horizon sooner than a receiver with a lower mask angle. As mentioned above, ideal geometry is one satellite directly overhead and the three other satellites evenly spaced around the horizon. If a receiver has a high mask angle, it will prematurely ignore the satellites providing the best constellation.

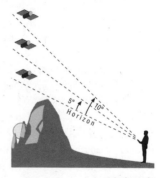

The GPS constellation is designed to provide coverage from at least four satellites at all times and at all places in the world. In populated areas, the constellation provides

Mask angle.

more than the minimum number of satellites. If your travels take you to remote areas of minimum satellite coverage, you will want the best antenna money can buy. If the receiver cannot detect all four satellites, it will either calculate a position using three satellites (2D mode), which is much less accurate, or it will suffer an outage.

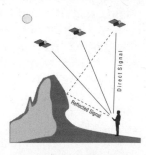

Reflected, or Multipath, Signals

In ideal conditions, the GPS satellite signals go directly from the satellite to the receiver. If a signal is reflected by something in the terrain, the receiver detects the direct signal and the reflected signal, so the signal arrives at the receiver by more than one path. In the figure, the signal from a satellite arrives at the receiver both directly and as a reflected signal from a nearby cliff. Presently, only survey grade (which means very expensive and usually heavy) receivers can detect and eliminate reflected signals. A consumer-grade receiver cannot tell the difference between the direct and reflected signal. If the receiver uses the reflected signal to make a position calculation, it will be inaccurate. As you use your receiver, be conscious of surroundings that may cause multipath error.

Reflected, or multipath, signal.

Elevation plotted over Distance.

Altitude Accuracy

The altitude (i.e., vertical position) provided by a GPS receiver is not as accurate as the horizontal position; however, WAAS provides altitude accurate to 6 m (19.7 ft.), which is pretty good. If you need more accurate altitude measurements, buy a good altimeter. Higher-end GPS receivers have built-in electronic altimeters and even track your altitude over time. Some wrist watches also have electronic altimeters.

Accuracy in 2D Mode

Receivers operate in two-dimensional (2D) and three-dimensional (3D) modes. Position

Tracking Elevation over Time with Barometer.

calculations in 3D mode are more accurate than in 2D mode. The 3D mode is the preferred mode of operation.

As mentioned above, a receiver uses signals from four satellites to calculate a 3D position (latitude, longitude and altitude). When the receiver locks onto only three satellites, it switches to 2D mode to try to provide a position even though it is less accurate. The receiver uses the signal from one satellite to synchronize time and the other two signals to calculate position. A 2D position calculation omits the altitude. The horizontal position accuracy in 2D mode ranges from 150 m to 1524 m (492 to 5000 ft.). A receiver working in 2D mode may not even get you within sight of your destination and you may have to lean a little more heavily on your manual navigation skills.

Dead Reckoning

A new integrated circuit developed by a company called SiRF Technology is designed to improve navigation accuracy in poor reception areas such as urban canyons (areas between man-made structures such as tall buildings where GPS signals may be blocked). The new circuit includes tiny accelerometers that provide dead reckoning when GPS signals are not available.

The circuit uses three-axis accelerometers to detect acceleration, and gyroscopes for measuring direction. When the circuit loses GPS signals, it keeps track of position using the accelerometers and gyroscopes. It tracks position using dead reckoning for as long as the GPS signals are blocked. Once the circuit detects GPS signals again, it uses them to calculate position. The circuit that includes the accelerometers and the gyroscopes is no larger than present integrated circuits, so GPS receivers that have this feature will be no bigger than today's receivers.

Future Satellite Networks

Currently, the United States is the only country with a fully operational Global Positioning System; however, other countries are launching their own programs and satellites. The Russian GLONASS system is being restored and updated to operational status with the goal of a fully operational system with global coverage by 2011.

Both China and the European Union have systems in development. The Chinese Beidou system currently has four satellites in orbit and is in an experimental phase. The final system, Beidou 2, is still in development and has no precise deadline for completion.

The European system, known as Galileo, received funding approval in May 2003 and is under development by the European Space Agency. One satellite has been launched for testing. The final system will include 30 satellites and should be in place by 2010. Galileo will provide two tiers of service. The first tier will be accessible without fee and will offer accuracy comparable to the U.S. GPS. The second tier will be a subscription service targeted at industry, airlines, shipping companies and developers. The predicted accuracy of the second tier service is a few centimeters.

Future GPS receivers may benefit from having multiple systems. The U.S. constellation has about 24 satellites. A receiver that detects and uses the signals from both the U.S. system and any other system will have access to between 42 and 52 satellites. Such a large constellation will avert outages and increase the likelihood of getting optimal satellite geometry. Further, it will provide a method for measuring ionospheric delay, because the various systems will be working on different frequencies. Receivers capable of detecting satellites from more than one system will improve accuracy, but they will be more expensive than current receivers when they are first introduced into the market.

3 About GPS Receivers

There are a lot of affordable GPS receivers on the market with features to meet every need. The importance of a feature depends on how you intend to use the receiver. The features and capabilities of hand-held GPS receivers are described in this and the next chapter to help you select the receiver most suited to your needs. Fortunately even receivers in the lower price ranges are loaded with features, so you probably will not need to make a choice between a feature and your pocketbook.

This chapter describes features common to all receivers, such as antenna sensitivity, map datum and grids. The next chapter describes features controlled by the user to operate the receiver.

Not all receivers have all the features described in these chapters, but you should be able to find one that has the combination of features you need.

Important GPS terms are also defined and explained in this section. The terms also appear in the Glossary at the back of the book.

From cars to trails, there is a receiver designed for every user.

Antennas

The antenna detects the signals sent from the satellites. It is the most important part of a receiver because if the antenna cannot detect the satellite signals, it cannot even begin to calculate your position. An antenna must have an unimpeded view of the sky to pick up the satellite signals. Antennas have only improved over the years, so just about any receiver has a more than adequate one.

There are two types of antennas: internal and external. Internal antennas are part of the receiver and generally cannot be disconnected or removed from the receiver. Most receivers have either a quadrifilar helix or patch (microstrip) antenna as an internal antenna. Quadrifilar helix antennas are generally rectangular or cylindrical in shape. Point a quadrifilar helix antenna toward the sky for best reception. Patch antennas are usually completely internal to the receiver, but some rotate away from the receiver. Hold the patch antenna parallel to the sky for best reception. Either type of antenna performs equally well in the field.

External Antennas

External antennas are separate from the receiver and connect to the receiver with a cable. An external antenna must be used whenever the receiver's antenna does not have a clear view of the sky. External antennas are passive or active. A passive antenna does not amplify the signals before sending them through the cable to the receiver. An active antenna amplifies the signals before sending them through the cable to the receiver. Passive antennas are less desirable than active ones because passive antennas may attenuate the signals and decrease accuracy. Active antennas use more power than passive antennas, so they are less desirable when the only power source is a battery.

Even though most GPS receivers can pick up the satellite signals through a windshield, external antennas make GPS receivers easier to use in a car, on a bicycle or even when backpacking. When a GPS receiver is mounted on a car dashboard, the driver must move toward the dash to see the receiver. A receiver connected to an external antenna may be moved to the driver, or preferably, to any other person in the vehicle. A GPS receiver attached to an external antenna can keep children in the back seat entertained for hours. External antennas are waterproof and can be used in any type of weather. They mount to the roof of a vehicle either magnetically or with suction cups.

A bicyclist may benefit from an external antenna by connecting the external antenna to their helmet or the outside of their backpack, so that the receiver can continuously receive the GPS signals and make a track log of the route pedaled while stowed in the backpack.

Navman windshield mount.

External antennas are small and can improve signal reception.

GPS bicycle mount.

Not all receivers are designed for external antennas. If you think you may want to use an external antenna, be sure to buy a model that accepts one. Today, many GPS antennas are available from vendors other than the GPS receiver manufacturer. It is easy to find a high quality, reasonably priced external antenna for just about any GPS receiver.

Accuracy

As discussed in the previous chapter, accuracy is an important factor in selecting a receiver. Preference should be given to WAAS-enabled receivers if WAAS or an equivalent system is available in your areas of travel.

Map Datum

A map datum is a reference point. Every location on a map is a known distance and height from the map datum for that map. A coordinate grid is a series of horizontal and vertical lines on a map that provide each location a unique coordinate number (e.g., latitude/longitude, easting/northing). A map can have numerous grids, but only one datum. When you use a GPS receiver, you need to tell the receiver the datum of the map you are using. Many receivers support hundreds of datum, so it is likely that such a receiver supports any datum you may ever need. If the only maps you ever intend to use are the electronic maps loaded into the receiver, the receiver already supports the datum used for the electronic maps and you do not need to worry about datum.

Most receivers for use in the U.S. support the two most common datum for North America: North American Datum 1927 (NAD 27) and World Geodetic System 1984 (WGS 84). If you travel or live internationally, you will need different datum, possibly a different one for each country. Some datum, such as OSGB, are unique to a grid. The OSGB datum and grid cover the U.K.

Before entering a coordinate into your receiver, be sure to set the receiver to the correct datum or there will be an error in the coordinate. For example, imagine you want to fly over Humphreys Peak, which is the highest point in the U.S. state of Arizona. From a USGS topographical map you measure the coordinate (latitude/longitude grid):

Humphreys Peak N 35° 20' 48", W 111° 40' 41"

When you enter the coordinate into the receiver, you do not notice that the Reunion datum is selected instead of NAD 27. In your plane, the receiver directs you to the point you stored, but it is not even close to the peak. When you check the receiver and change the datum to NAD 27, the coordinate you stored changes to:

Humphreys Peak N 35° 19' 55.8", W 111° 40' 21"

Entering the coordinate with the wrong datum resulted in a position error of 1.7 km (1.05 mi.). The Reunion datum is an extreme example. The differences between some datum are small, but all the same, you do

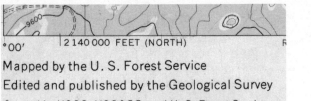

°00'   2 140 000 FEET (NORTH)   R

Mapped by the U. S. Forest Service

Edited and published by the Geological Survey

Control by USGS, USC&GS, and U. S. Forest Service

Topography by photogrammetric methods from aerial photographs taken 1967.   Field checked 1972

Projection: Utah coordinate system north zone (Lambert conformal conic)
10,000-foot grid ticks based on Utah coordinate system, north and central zones
1000-meter Universal Transverse Mercator grid ticks, zone 12, shown in blue.   1927 North American datum

Most maps contain a reference to UTM zone number and datum.

not want an incorrect datum to be the source of position error. Be sure to set the datum correctly before you enter any coordinates.

There are hundreds of map datum. Some examples are:

WGS 84: World Geodetic System 1984
A datum for the whole world as used by the GPS.

NAD 27: North American Datum 1927
Used at present by older Canadian and U.S. maps.

NAD 83: North American Datum 1983
Used by newer Canadian maps and U.S. maps in the future.

OSGB: Ordnance Survey Great Britain
Great Britain, Scotland, Isle of Man....
Note there is also a grid by the same name – do not get confused.

Geodetic Datum 1949
New Zealand

Selecting the NAD27 CONUS map datum.

Built-in and Downloadable Maps

A built-in map is an electronic map that is stored in the receiver and displayed on the receiver's screen. Built-in maps are what make GPS receivers fun. Although a GPS receiver makes using paper maps a lot easier, maps displayed on the receiver's screen along with your present position and route to a desired waypoint make GPS receivers a powerful navigation tool.

Built-in maps include base maps and detailed maps. A base map usually has all major highways and some major surface streets for a given area, such as the U.S., Canada, Europe, South Africa or Australia. Many receivers have a base map. A detailed map augments the base map and contains minor streets and points of interest (POI) such as hotels, restaurants, hospitals, airports, museums and gas stations. Receivers with detailed maps can search for locations by address, intersection,

Display of built-in map.

street name and location name. Base maps are generally stored in the receiver's memory. Detailed maps generally are downloaded onto a memory card that is inserted into the receiver (e.g., an SD card).

All receivers designed for cars and some designed for backcountry use provide point-by-point directions that follow the streets and highways shown in the built-in map. Downloadable topographical maps are also available in 1:24,000 and 1:100,000 scales. Many of the downloadable topographical maps display trailheads and other useful information.

Memory

If you purchase a receiver with the intent of using it with downloadable maps (detailed map), choose one that accepts some form of memory card, such as SD cards. Downloadable maps can require a lot of memory, but fortunately memory cards are high capacity and very affordable. Instead of wondering if your receiver has enough built-in memory to support your electronic map needs, simply purchase an SD card that is big enough to handle all your foreseeable electronic map needs. If your map storage needs increase, simply purchase a higher-capacity card.

The amount of memory required to store all the road maps of the greater Los Angeles metro area is 21 megabytes. The entire road map for North America can be fit onto a microSD card that holds 2 gigabytes.

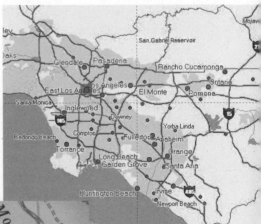

above:
Los Angeles and surrounding areas.

left:
2 GB microSD card with North American Road Atlas.

Coordinate Grids

As mentioned above, a coordinate grid is a pattern of horizontal and vertical lines drawn on a map to uniquely describe every point. The grid identifies a place on a map using a combination of letters and numbers called a coordinate. Different locations on the same map cannot have the same coordinate. GPS receivers display coordinates, so once a receiver locks onto the satellite signals, the letters and numbers displayed by the receiver are your position coordinate. When considering a GPS receiver for use with a paper map, it is important to buy one that supports the grid used by the map.

At about this moment, you are probably thinking you can skip this section and all other sections that deal with using paper maps and reading coordinates. After all, you have seen or own a receiver with electronic maps that display your current position as a dot on the screen and not as a coordinate. You also know that you can navigate using the electronic maps displayed by the receiver without knowing your coordinate. That knowledge leads you to conclude that grids are unimportant. You are

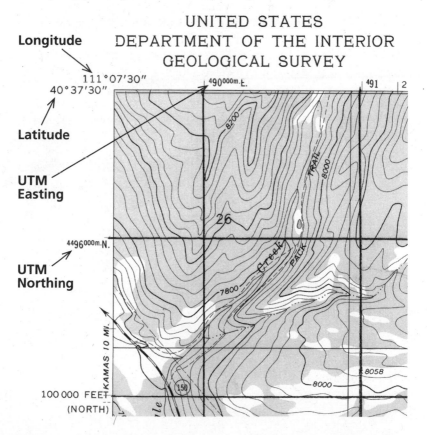

29

partially correct, but the part in which you are in error makes understanding grids even more important.

Electronic maps have only improved over the years. They are available for more areas than ever before. Many sources offer electronic maps of wilderness areas, so you can go off road without ever being forced to read a coordinate. Until you want to go somewhere that the electronic maps do not cover or the electronic maps are inferior to paper maps, your motivation to learn about coordinates will be low. Many urban travelers and outdoor adventurers in well-traveled areas can rely exclusively on electronic maps. However, the day has still not arrived when every square inch of any country, including the U.S., is entirely covered by electronic maps with sufficient detail to completely eliminate paper maps. Coordinates are necessary to those who leave the beaten path or crave the rich detail of a large-scale map as opposed to the tiny portion of a map shown on a relatively minuscule receiver screen.

For those who know or even suspect that electronic maps cannot guide them to their destination, look closely at the grids supported by your GPS receiver. Do not buy a receiver that does not support at least the Universal Transverse Mercator (UTM) and latitude/longitude grids, because together they cover most of the paper maps used in the world. Some receivers only support UTM and latitude/longitude, which is enough for most users, but most receivers support additional grids. Some grids pertain to specific countries, such as the Ordnance Survey grid of Great Britain. If you plan to use your receiver in Great Britain, it is best if your receiver supports the Ordnance Survey grid, because that is the grid used in a wide variety of British maps.

Some Popular Grids

- Universal Transverse Mercator (UTM)
- Latitude/Longitude
- British Grid (OSGB)
- Maidenhead
- Military Grid Reference System (MGRS)
- Universal Polar Stereographic (UPS)

Selecting a coordinate grid.

The most common grids, UTM and latitude/longitude, are thoroughly explained and demonstrated in Chapters 5–6 and 7–8 respectively. The other grids are briefly described here and more information is provided Chapter 13.

OGSB

Used by the excellent Ordnance Survey maps of Great Britain.

MGRS

The grid used by the U.S. military. It is based on the UTM grid, but it replaces some numbers with letters. Until the advent of electronic map databases, the MGRS was not readily accessible to civilians. Now, anyone can print a map with the MGRS grid. See Chapter 13 for more information.

Maidenhead

The grid system used by amateur radio operators.

UPS

Developed to cover the Arctic and Antarctic. Similar to the UTM grid.

All receivers can switch from one grid to another, and some display two coordinate grids simultaneously. Other receivers make it easy to switch between coordinate grids. If you know you will need to use coordinates in two different grids, buy a receiver that allows conversion between the two to be done with as few button presses as possible.

Below are some examples of coordinates of the same place in different grids. Note that the OSGB grid is valid only in Great Britain.

Calgary, Alberta, Canada

11 U 703421m.E. 5662738m.N.	UTM
N 51° 4' 55.2", W 114° 5' 44.6"	Lat/Long
11 U QG 03421 62738	MGRS
DO21WB	Maidenhead

New York, New York, USA

18 T 583926m.E. 4507327m.N.	UTM
N 40° 42' 50.9", W 74° 0' 23.0"	Lat/Long
18 T WA 83926 07287	MGRS
FN20XR	Maidenhead

Harrogate, England

SE 31000 55000	OSGB
30 U 596556m.E. 5982957m.N.	UTM
N 53° 59' 23.7", W 1° 31' 37.7"	Lat/Long
18 T WE 96556 83957	MGRS
IO93FX	Maidenhead

Subsequent chapters explain more about the grids and will help you decide which one is best for your navigation needs.

Computer/PDA Interface

Computers and Personal Digital Assistants (PDAs) are useful pieces of navigation equipment. Although you will not take your desktop computer into the field for navigation, you can use it to store waypoints and other information after a trip and to download maps or waypoints before a trip. A laptop can be used to navigate in a vehicle. The laptop may receive position information from a GPS receiver and show position and routes on its screen. A PDA connected to a GPS receiver can go into the field, whether urban or backcountry.

If a computer or PDA will be part of your navigation plan, be sure your receiver has a computer interface. Most receivers are capable of two-way communication with a computer, which means that information transfers from the computer to the receiver and vice versa. Do not buy a receiver that only supports one-way communication. You do not need to worry about the computer not speaking the receiver's language, because there are standards set by the National Maritime Electronics Association (NMEA) that are adhered to by both software and hardware manufacturers. The current NMEA protocol is 183 version 3.01. Be sure your receiver supports it.

NMEA 183 version 3.01

As described above, NMEA is an industry-standard language. Some receivers also support earlier versions such as 2.0 or 1.5. Any version is capable of working with other types of equipment. The NMEA protocol is used to communicate with moving maps, chart plotters, automatic pilots and other types of equipment that need position information to do their job.

RTCM SC-104

RTCM SC-104 is the industry-standard format for Differential GPS correction information. WAAS make the RTCM format unimportant for most users. It may be important to commercial users, depending on their application.

Text

Some receivers are capable of sending out text (ASCII) characters. Text output is not governed by an industry standard, so the information provided will depend on the manufacturer.

Selecting NMEA for the computer interface.

Channels

Modern receivers can have up to 20 parallel channels. Most receivers have 16 parallel channels that permit the receiver to simultaneously track up to 16 satellites. Receivers with more channels can lock onto more satellites. Tracking many satellites simultaneously enables the receiver to always select the satellites that provide the best geometry, thereby increasing accuracy.

The term "parallel channels" means that all the satellites are simultaneously and continuously tracked. In operation, the satellite signals enter the antenna, then go to the channels. One channel locks onto the first satellite in view, another locks onto the next satellite and so forth until every satellite visible in the sky is assigned a channel. The channel continuously tracks its assigned satellite as long as it is in view. The information from the channel goes to the navigation processor, where the satellites that provide the best DOP values are used to calculate position. Rarely does an area on the globe get coverage from more than 10 satellites, so 16 parallel channels surpass normal needs.

Parallel channels ensure that the navigation processor always has the most up-to-date information to make position calculations. If you are walking past a big tree in the woods that suddenly blocks a satellite, the navigation processor immediately uses the information from the next best satellite to calculate your position.

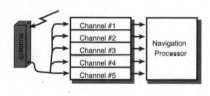

Multiple Channel System

Multiple channels simultaneously track multiple satellites.

Compass

Do not throw your compass away, unless of course you purchase a receiver with a built-in electronic compass. A receiver, without a built-in compass, is not at all like a compass. All receivers report your direction of travel by calculating the direction between your current position and where you were a few seconds ago. When you stop, the receiver can no longer calculate your direction of travel. It usually displays your last direction of travel until you start moving again. If you stand and slowly turn around in a circle, the bearing displayed by the receiver will not change because you have not moved far enough for the receiver to calculate a direction.

A compass uses the earth's magnetic field to detect your bearing relative to the magnetic pole. If you hold the compass and turn around slowly, the compass needle moves and continuously points to the magnetic pole. An electronic compass built into a receiver also detects the earth's magnetic field, so it can determine your direction even when you are not moving. The great thing about a built-in compass is that the receiver can automatically compensate for declination, something you have to do manually with a regular compass.

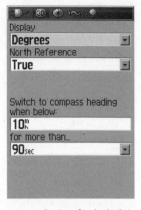

Settings for the built-in electronic compass.

Most receivers allow the user to specify when the electronic compass reading will be used, as opposed to the receiver's calculated bearing. Generally, when you are moving faster than a minimum speed, the bearing calculated by the receiver will be used. When moving slower than the minimum speed, the bearing detected by the electronic compass will be used. As shown in the screen shot above, the user may set the minimum speed.

Electronic compasses use very little power. Some receivers permit you to turn the GPS circuits off while leaving the compass on to provide bearings. Battery life is extended by many hours when only the electronic compass is used. Using the electronic compass is much like using a regular compass in that the instrument must be held parallel to the ground when taking a reading.

The compass orients you along a straight-line course to your destination, usually called the bearing or heading. A GPS electronic compass may also display a "Course Pointer." This feature is discussed in more detail in the section on Course Deviation Indicators. By switching the electronic compass to the course pointer mode, the compass guides you to follow a course between two waypoints. If you stray from the course, the arrow on the compass will continue to point to the direction destination and not your current direction of travel.

Hold the receiver level, just like a regular compass when taking a reading.

Altimeter

Some receivers include an electronic altimeter. Although receivers use the satellite signals to calculate altitude, a built-in altimeter is more accurate, especially in areas not covered by WAAS.

Built-in altimeters work off of barometric pressure, which is the pressure of the surrounding air, or atmospheric pressure. The atmospheric pressure at sea level is greater than the atmospheric pressure at the top of Mount Everest because as you climb higher in altitude, the amount of atmosphere above you decreases. An altimeter must know its starting elevation. Then, as the barometric pressure varies, the altimeter converts the changes in pressure into changes in altitude readings. Unfortunately, a change in weather also brings a change in atmospheric pressure, which affects the accuracy of a barometric altimeter.

Enabling the Course Pointer.

The units of atmospheric pressure are inches of mercury, millibars or kilopascals. Be sure the receiver supports the unit of atmospheric pressure used in your area. Most receivers allow the altimeter to be used in one of two ways:

- User calibrated
- Automatically calibrated

In the user-calibrated mode, the user must specify the initial elevation or atmospheric pressure and must periodically, depending on changes in weather, recalibrate the altimeter. Automatic calibration uses the elevation calculated by the GPS receiver as the starting elevation for the electronic altimeter. Automatic calibration also periodically recalibrates, using the current GPS-calculated elevation.

Barometric pressure as a function of time.

If the weather is not changing and you have an accurate starting elevation, the altimeter elevation will be more accurate than the elevation calculated by the receiver, so set the receiver to the user-calibrated mode. If the weather is changing or you do not know the starting elevation, use the automatic calibration mode. The automatic calibration mode enables the altimeter to improve altitude readings at least until the altitude is again automatically recalibrated.

Power Sources
Batteries
Most receivers use conventional batteries. However, some newer receivers come with non-standard rechargeable batteries similar to those found in cellular phones. How you use your receiver will depend on whether you

want a receiver that accepts conventional batteries or a non-standard rechargeable battery. If you use your receiver in the backcountry, you want a receiver that uses conventional batteries, so you can carry fresh batteries for exchanging during the trip. If you use your GPS receiver in a vehicle that has a cigarette lighter, you can recharge your receiver at any time, so a receiver that uses non-standard rechargeable batteries is fine.

There are a few things you can do to conserve energy regardless of the type of battery used. If you are on foot, use the receiver only when you need to locate your position on either a paper map or the built-in electronic map. Keeping the receiver off most of the time clearly saves on batteries. Another method of saving battery power is to limit use at night. Most receivers use a type of screen called a liquid crystal display (LCD). If you want to see the screen at night, you have to turn on a small light bulb behind the screen called the "backlight." The backlight quickly drains the batteries, so use it sparingly.

Changing the Batteries

Do not worry about losing data when you change the batteries. Most receivers have a small internal battery that maintains the data even when the power source is removed for an extended period of time. Newer receivers use a special type of memory called "flash memory" that never loses its data regardless of how long the batteries have been dead.

External Power

If external power is available, such as a cigarette lighter in a vehicle, use it. Using an external source permits you to leave the receiver and backlight on continuously. It will save you the expense of replacing the batteries and will allow the receiver to operate as a true navigational aid.

Multitone or Color LCD Screens

Most receivers now have a multitone (e.g., grayscale) screen, which means that instead of having only black and white, they have black, white and shades of gray. Some receivers have full-color screens, which makes them much easier to read and understand; however, color can add a lot to the price of a receiver. Furthermore, multitone screens are easier to read in direct sunlight.

Some receivers allow the user to determine whether land or water will have a dark shade of

Water is a lighter color and land is a darker color.

gray for its background. If you travel primarily on land you do not want the land to have a dark background, because it obscures the marked waypoints. If you travel primarily on water, you want water to have a light background because you want to see the details of the water.

Land is a lighter color and water is a darker color.

Mounting

All manufacturers sell hardware to mount their receivers. It has been mentioned before that most receivers can be set on a car's dash to pick up the satellite signals through the windshield. Mounting hardware fixes the receiver to the vehicle so that sudden turns or stops do not send it flying. Mount the receiver within reach of the driver or pilot so they can press the buttons and see the receiver's screen. A receiver mounted deeper inside the vehicle makes it difficult for the receiver to pick up the satellite signals without an external antenna. Both mounting hardware and an external antenna are worthwhile investments. Receivers may also be mounted on ATVs, bicycles, kayaks, motorcycles and even runners' wrists.

Weight

Advances in electronics have produced several lightweight GPS receivers. Most modern receivers, even the more powerful ones, run on 2 AA or AAA batteries, weigh between 3 and 7 ounces, and easily fit in your pocket. Generally, the weight of a receiver is tied to the size of its screen and the number of batteries it uses. If you are on foot, you will want a smaller, lighter receiver that uses batteries sparingly. If you are in a vehicle, you will want the largest screen possible without regard to weight or power consumption. If you travel in a vehicle most of the time, but occasionally need a receiver in the field, you can find higher performance hand-held devices that provide built-in maps that calculate routes that follow the roads.

Modern receivers are small and lightweight.

Data Entry Keys

The data entry keys on receivers range from buttons on the face, buttons on the side, touch screens and mouse-like nubs similar to those on portable computers. The common characteristic is that they are small, so if you have to use a receiver while wearing gloves, use a pencil or something else to press the buttons, because your gloved hand will be too big to do it.

Some receivers may be set to beep each time a key is pressed. If you use a receiver in the dark, the backlight illuminates the screen but not the keys. If the receiver beeps with every button press, you can be sure the key was actually pressed.

Most receivers have buttons; some provide a touch screen.

Lanyard Attachment

If you plan to use your receiver while ski touring, or while backpacking with two poles, make sure a lanyard can be attached to the receiver so you can hang it around your neck. Some receivers are now very small, so a lanyard will help prevent losing it. Some receivers come with points to attach belt clips so you can easily attach your receiver to your belt or backpack for travel.

4 Common Functions of GPS Receivers

The functions described in this chapter may be found in many receivers available on the market. Different receiver brands may use different names to describe the function. Understanding the functions available will help you understand which functions are important to your application.

Waypoints

The coordinate of a location is called a waypoint or landmark. Waypoints are fundamental to GPS technology. Navigation with a GPS receiver starts by providing a destination waypoint to the receiver. You may provide waypoints to the receiver in numerous ways such as typing the coordinate and name into the receiver's keypad, selecting a point from the receiver's Map screen, selecting a point of interest (POI) from a list on the receiver, or selecting a waypoint from a map on your computer and transferring the waypoint to your receiver. Once the receiver has the destination waypoint, it determines your present position, calculates the distance, direction and route to the destination waypoint and directs you to follow the route

Because waypoints are the foundation of GPS navigation, you want your receiver to be able to store as many waypoints as possible. Most receivers store between 500 and 1000 waypoints. POIs are not stored in the same memory as user-defined waypoints, so a receiver may provide thousands of POIs and still permit you to store hundreds of waypoints. In receivers that provide POIs, the POIs may be selected from lists of categorized POIs. Destinations may also be specified by street address.

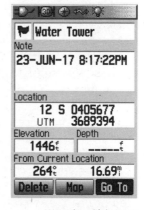

Waypoint with its name, coordinate and elevation.

Any receiver will store enough waypoints for at least one trip. If you want to track places you have been over a long period of time, you will need to transfer waypoints from your receiver to a computer for long-term storage or the receiver will soon run out of memory for waypoint storage. If you use a computer as part of your navigational equipment, you can manage all your waypoints on your computer and transfer the waypoints you need to the receiver before the trip. Most users will find 500 waypoints to be more than adequate and transferring waypoints to a computer an inconvenience. Users who mark terrain for purposes other than navigation (e.g., occupational or vocational users), such as marking property boundaries, mining claims, geocache locations or accident sites, will find 500 waypoints to be restrictive and will need to use a computer to store and manage their waypoints. Software for waypoint management is discussed in Chapter 9.

Each waypoint consists of a coordinate and a name entered by the user to describe the waypoint. Some receivers limit names to ten characters, so creating a descriptive name may be challenging. Some receivers, especially those designed for outdoor use, provide symbols to mark types of waypoints (e.g., home, dock, trailhead). Some receivers provide a line for entering text to describe the waypoint. However, entering text information via the receiver buttons is tedious, so if you use the descriptive line, use a computer to enter the information.

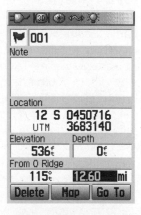

When using a receiver in the field, if you do not want to enter a name, most receivers will automatically generate one for you. For example, the first time you mark a waypoint, the receiver will call it 001, the next waypoint will be called 002, and so forth. You can either rename the waypoints later or keep the receiver-generated names, since their position is displayed in the receiver's Map screen.

Automatically generated name.

Most receivers also allow you to choose a symbol that appears on the Map screen. There is an abundance of symbols to choose from, with categories such as markers, outdoors, marine, civil, transportation, signs and points of interest. Some receivers even let you create your own custom symbols.

All waypoint information – the name, the coordinate and the comment – are stored in the receiver's memory. Provided the batteries are not dead, information in the memory is not lost when the receiver is turned off. Some receivers have small backup batteries or use flash memory (the same type of memory as in cell phones) to preserve waypoints even if the batteries go dead or are removed.

Symbols can distinguish waypoints on the map.

Waypoint Manipulation

Because waypoints are fundamental to GPS navigation, your receiver must make it convenient and easy to enter, retrieve and modify them. If the receiver makes you press a lot of buttons to manipulate waypoints, you will not enjoy using it and it will collect dust on the shelf. With hand-held receivers designed for the outdoors, you want to access and store waypoints with a minimum number of button presses. Listed below are the convenient ways many receivers allow users to access the waypoints stored in memory.

- Alphabetized lists with prompting
- Waypoints listed by proximity to present position or a position on a map
- Selection from the receiver's Map screen
- Waypoints in chronological order

All receivers display the waypoints stored by the user. More powerful receivers provides Points of Interest (POI), cities, freeway exits and address look-up of locations. All waypoints, whether user-stored or pre-programmed POI, are accessed by one of the above methods.

For the detailed descriptions below, the user-programmed waypoints include the waypoint names Boat In, Boat Out, Bridge, Camp 1, Camp 2, Home, Lake In, Lake Out, River Entrance and School. Below is how a typical receiver accesses these waypoints using the methods listed above.

Alphabetized List

When searching for user-stored waypoints, receivers generally display an alphabetized list of the waypoints in memory. You can scroll down the list until you find the desired waypoint or you can type in the first letters of the waypoint name and have the receiver search for any names that start with those letters. Scrolling down a list of 500 waypoints is slow, so searching by entering the first few letters of the name is much easier. The receiver displays the waypoints that match the letters as a list, and prompts the user to select a waypoint. Unfortunately, receivers are not like a typewriter where you can randomly select

A list of waypoints.

any letter of the alphabet. There are two common ways receivers enable the user to access the alphabet: sequentially and by a table.

Sequential selection means you start at the beginning of the alphabet and cycle sequentially through the letters, either in the forward or reverse direction, until you reach the desired letter. For example, you press a button and the receiver displays the letter "A." Press the button again to get "B," then "C," and so forth, until you reach the desired letter. Cycling serially through the alphabet is slow, so the receiver speeds up the search for waypoints by prompting with the names of the waypoints that match the letters entered. Generally, you only need enter the first few letters of a waypoint name before it is displayed for selection.

For example, if you want to select the waypoint Home, you cycle through the alphabet for the first letter. When "B" is displayed as the first letter, the receiver shows the name "Boat In" on the screen. It is prompting you to select the first waypoint that begins with a "B." If you want to see the Boat In waypoint, you would press enter, but as you are looking for Home, you press the button again, the first letter changes to "C," at which point the receiver prompts you again with the waypoint named "Camp 1." You continue scrolling until you reach "H," at which time the waypoint "Home" is displayed and you select it by pressing Enter.

Table selection provides faster access to the alphabet. As shown in the figure, an alphabet table displays the alphabet in rows and columns. To select any letter, you press the up, down, left or right buttons to move among rows and columns of the table. Selection of any letter from most tables can be done with at most eight keystrokes. The receiver displays waypoint names as you enter the letters, so when you select "H" the receiver displays the waypoints that start with "H." You can enter the second letter, "O," or simply select Home from the alphabetized list. An alphabetized list combined with an alphabet table provides fast, easy access to stored waypoints.

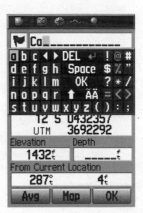

Selecting Camp waypoint using a table.

Nearest-Waypoint List

If you have a receiver that stores lots of waypoints or thousands (possibly millions) of POIs, displaying waypoints by proximity to your current position provides easy access to the ones of most immediate interest. The waypoints closest to your present position are displayed at the top of the list. Some receivers also provide waypoints by proximity to a location selected on the map, so you can look for POIs near a location even before you arrive.

Listing waypoints by proximity saves lots of time. Furthermore, POI waypoints are generally divided into categories such as entertainment, food, shopping, marine, fuel services, lodging, banks, recreation, attractions, hospitals and transportation, so you can save even more time by selecting an appropriate category before searching the proximity list.

Hospitals by proximity, from closest to farthest away.

Waypoint from Built-in Map

A receiver that has a built-in map allows waypoints to be selected directly from the Map screen. Any waypoint shown on the map may be selected by moving the cursor, usually an arrow, to the waypoint. When the cursor touches the waypoint, you can select the waypoint and its information is displayed on the screen. It is also possible to select user waypoints from the Map screen. The Map screen is a visual version of the nearest-waypoint list.

Waypoint by Time of Selection

Some receivers list waypoints according to the time when the waypoint was last selected. The most recently selected waypoints are at the top of the list. This feature is useful, for example, when you selected a waypoint from the map or from a POI list but did not store it as a user waypoint. The list shows previously selected waypoints in reverse chronology of selection.

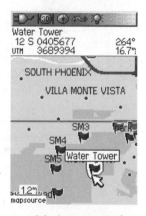

Selecting a waypoint from the Map screen.

Proximity or Dangerous-Waypoint List

Some receivers provide a proximity list that warns you when you approach hazardous locations. The coordinates of the dangerous areas must be entered as waypoints and placed on the proximity list. The list allows you to specify how close you can get to the object before the receiver sounds a warning. If you receiver's units were set to statute miles, the list shown would keep you 1.5 mi. (2.4 km) away from Boat Out, 2 mi. (3.2 km) from Bridge and 5 mi. (8 km) from River Entrance.

Setting proximity alarms.

Goto Function

The true power and utility of a GPS receiver is summed up in the word Goto. All receivers are capable of leading you to a destination waypoint. You simply select a waypoint (e.g., user or POI) or select a location from the built-in map and tell the receiver to guide you there. The power to Goto any place on earth is fundamental to GPS navigation.

Select user waypoints or POIs.

The receiver guides you to the destination waypoint using a steering screen. There are several different versions of steering screens, as described later in this chapter, but they all fundamentally do the same thing. A steering screen points the direction from your current position to the destination waypoint. The receiver not only indicates direction to the destination, but also tells you if you are off course, your speed, when you should arrive, and other useful information that is described in the section titled Navigational Statistics later in this chapter. Some receivers have a separate Goto button, while others activate Goto by finding a waypoint from a list. Once you select your destination, follow the receiver's directions displayed on the steering screen.

On the Compass steering screen shown below, your direction of travel is represented by the vertical line at the top of the compass at about 105°. The direction you need to travel to get to the next waypoint is indicated by the arrow, which points due west (270°). To travel toward the destination, you need to turn to the right 165°, which means you are currently headed in nearly the opposite direction.

A receiver may lead you to the destination along a straight line or it may plan a route that follows the roads displayed on the built-in map. Many middle-range and lower-end units display built-in maps, but cannot calculate routes that follow the roads. All vehicle-related receivers and many higher-end hand-held receivers calculate routes that follow roads. A huge advantage of a receiver that routes along roads is that it can calculate an accurate time of arrival for road travel. Some hand-held receivers permit the user to selected whether routing follows the roads of the built-in map or whether routes between waypoints are a straight line (off road).

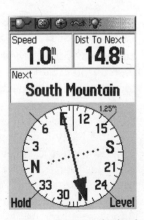

Turn around to head toward South Mountain.

If you primarily use your receiver for back-country adventures, but would still like to use it on the road occasionally, purchase a high-end model that allows you to select whether a route follows the roads or is straight line. If your receiver provides only a straight line to a destination, routes may be used to break a trip into many segments, with straight lines between each segment, but entering the waypoints to form such routes is time consuming.

Following roads or direct course.

Routes

A route is a list of waypoints that describe the path you will travel. It is like the Goto function but it leads you to many points sequentially instead of to a single waypoint.

The route function is important because it enables the receiver to guide you from the first point in the route to each successive waypoint until you reach your final destination. Receivers that follow roads automatically form a route that directs you from one intermediate point to the next until you arrive at the final destination. The route function is similar to the Goto function because it allows you to specify where you want to go, but it is more powerful because you can also choose the path you take.

The Route function is more automatic than Goto. When you reach one waypoint on a route, the receiver automatically guides you to

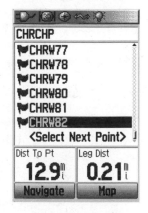

Route CHRCHP has 82 way-points and is 12.9 miles long.

the next waypoint without your having to touch a single button. With the Goto function, you must manually select the next waypoint before you start out again. Most receivers, except for the most basic, have at least one route. The route function should have the following capabilities:

- Off road: 50–100 waypoints per route
- On road: automatic route calculation that follows roads; automatic recalculation
- Automatic route reversal
- Display navigation information between points

Routes may be assembled manually or calculated automatically. Vehicle-related receivers automatically calculate routes along roads. The user does not have to store any intermediate points. The user simply provides the destination and the receiver does the rest. If the user strays off the route calculated by the receiver, the receiver calculates a new route. The most useful vehicle-related receivers provide voice prompts to guide you along the route.

Manually assembled routes are formed using waypoints stored in memory or locations selected from the built-in map. You form the route by placing the waypoints on a route list in the order they must be traversed. Once the route is formed, you activate the route function and the receiver guides you from the first waypoint to the next and so on until you reach your destination.

Routes may also be formed using a computer and then transferred to the receiver. The number of waypoints in a route is limited, so whether you form the route manually or transfer it from a computer, the number of waypoints cannot exceed the limit for the receiver. Generally, the number of waypoints permitted per route is enough to fill any need.

Automatic route reversal means the receiver makes the destination the starting point and the starting point the destination, and puts all the intervening waypoints in the correct reverse order. Automatic route reversal means you do not have to manually form another route when you want to return. Once the route is reversed and activated, the receiver once again points the way from one waypoint to the next until you arrive at your original starting point.

Most receivers calculate and display the bearing and distance between the waypoints that form the route to give you a rough idea of the trip's total length.

Routing along Roads

Receivers that are capable of calculating routes that follow roads may have several options that make the routing feature more useful for the user. Automatic route recalculation can be enabled so that when you leave the indicated course the receiver will automatically recalculate a new route to guide you from your present position to the destination. Automatic recalculation is vital because you are bound to occasionally miss a turn indicated on the route. Most receivers can recalculate the route so quickly that you do not get too far before getting new instructions on how to proceed.

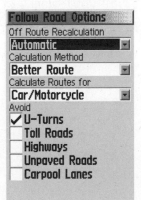

Options for routes that follow roads of a built-in map.

You can also program the receiver to avoid certain kinds of road features such as highways, toll roads and U-turns. Some vehicle GPS receivers can receive traffic updates via an FM radio attached to the receiver. The radio receives traffic information and identifies slow traffic locations to the receiver. The receiver uses its routing capabilities to guide you around traffic.

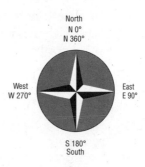

Bearing (Azimuth)

The term "bearing" as it is used in this book is more correctly called an azimuth. Look for both words in the Glossary for a deeper explanation. Most people use the word bearing as the compass direction between your present position and your destination. The bearings of the cardinal compass directions (East, West, North and South) are shown in the figure. If you travel due east, your bearing is 90°.

The bearings between the waypoints Mine, Camp and Town are:

Start	Destination	Bearing
Mine	Camp	30°
Mine	Town	295°
Camp	Town	240°
Camp	Mine	210°
Town	Camp	60°
Town	Mine	115°

Some receivers use the word "bearing" as the direction you must go to get to a waypoint, while the word heading denotes your current direction of travel.

Receivers can report the bearing between your present position and any other waypoint or between any two waypoints. Bearings can be relative to the North Pole or the magnetic pole, so be sure you know which one is selected before you use a bearing. If the true north mode is set, the receiver automatically compensates for declination anywhere in the world. The difference between true and magnetic bearing is explained in the section titled North Settings.

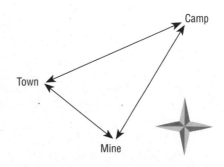

North Settings

There is more than one north, and when the receiver reports a bearing, you need to know which north it references. A receiver provides some or all of the following modes:

- True north
- Magnetic north
- User-defined north
- Grid north

Before explaining the various types of north, it is important to understand the difference between the two most important norths: true north and magnetic north. True north is the direction to the North Pole, which is one end of the axis through the center of the earth. Magnetic north points to the magnetic pole, which is southeast of the North Pole and northwest of Bathurst Island in northern Canada. Magnetic declination is the difference, in degrees or mils, between the North Pole and the magnetic pole from your position. The figure to the right shows how declination is measured. It is expressed as an east or west direction depending on whether the magnetic pole is to the right or left of your present position. The larger figure on the next page shows declination throughout the world.

Magnetic declination in North America.

In **true north** mode, the receiver shows all bearings as referenced to the North Pole. When the receiver reports a direction of 0°, you are headed directly toward the North Pole. A bearing of 180° takes you directly to the South Pole. Maps are oriented to the North Pole, so set the receiver to the true north mode when using it with a map. A bearing is converted to magnetic simply by switching the receiver to the magnetic north mode or by using the manual conversion described below.

To manually convert a **true north** bearing (map bearing) to a magnetic bearing (compass bearing), remember the following phrase:

East is Least, West is Best

This phrase means that you subtract east declinations and add west declinations to convert from true north to magnetic north. Here are some examples:

- Map bearing   = **50°**,   Declination = **East 12°**
  Compass bearing = 50° – 12°   = **38°**

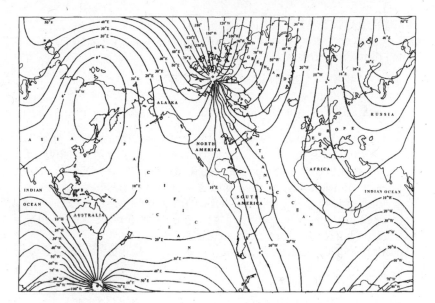

Magnetic declination around the world.

- Map bearing   = **276°**,   Declination = **West 17°**
  Compass bearing = 276° + 17°   = **293°**

Magnetic north mode references all bearings to the magnetic pole. When the receiver is in the magnetic north mode, all bearings directly relate to a compass. If the receiver says the bearing to the destination is 109°, set your compass to 109 to travel to that destination. To convert a magnetic bearing to a true north bearing, simply change the receiver mode to the true north mode or use the inverse of the manual procedure described above. Receivers are programmed with declination information to perform the conversion between true north and magnetic north.

User-defined north allows the user, not the receiver, to specify declination.

Grid north is the direction to which the grid on the map is aligned. In most situations, grid north is the same or almost the same as true north. Some receivers allow you to enter the grid declination if you are using a map that does not have the grid aligned with true north. Usually the difference between grid north and true north is so small that it can be ignored without consequence.

UTM GRID AND 1974 MAGNETIC NORTH
DECLINATION AT CENTER OF SHEET

Declination diagram from a USGS 7.5-minute series map. The magnetic declination is E 13.5°. The difference between true north and grid north is W 14 mils.

49

Data Formats

Most receivers report speed, distance, altitude, CrossTrack Error, etc., in three different formats:

- Metric: kilometer, kilometers/hour
- Nautical: nautical mile, knots
- Statute: mile, miles/hour

It is important that a receiver be able to display information in all the above formats, because various coordinate systems are best suited for use with a particular unit. For example, UTM, MGRS and OSGB are best used with metric units, as their grid is a kilometer-based grid. The latitude/longitude grid is based on nautical units, which are used on many marine charts; however, most land adventurers in the U.S. prefer to use statute units. Some receivers are capable of reporting speed, distance and altitude in different units, which means you can mix and match units to your liking.

Receivers provide two units for bearings:

- Degrees
- Mils

Bearings are more widely reported in degrees, but it is nice to have a receiver that can do the conversion between the two. If the receiver has a built-in altimeter, the unit selected for elevation is the unit that must be used for calibrating the altimeter.

Selecting the units to be displayed by the receiver.

Navigational Statistics

If you are on foot, it is nice to know the direction and distance to your destination, but information like speed, estimated time of arrival and other navigation statistics are not as important, because you are moving slowly and you probably do not keep the receiver on all the time. The situation changes when you are in a car, boat, plane, snowmobile or any other vehicle where you leave the receiver on all the time and get continuous information as to your location and travel toward the destination. Your specific use of a GPS receiver will determine which navigation statistics are most useful to you. Read the descriptions below of each statistic, and then look for the receiver that provides the information you need.

Navigational statistics available on most GPS receivers include:

- Distance
- Speed
- Desired course (bearing)
- Current course (heading)
- CrossTrack Error (XTE) aka Course Deviation Indicator (CDI)
- Course to Steer (CTS) aka To Course
- Turn (TRN)
- Estimated Time en Route (ETE)
- Estimated Time of Arrival (ETA)
- Time
- Glide Ratio

A first configuration of navigational statistics displayed.

GPS receivers designed for vehicular road use also provide:

- Turn-by-turn Instructions

A receiver may also allow you to select which navigational statistics are displayed on the screen. For example, the first screen displays time-related statistics. The second screen displays speed and bearing. Each section of the screen may be programmed to display any desired navigational statistic, whether it be time, distances, bearing, heading, etc. The data displayed may be changed by selecting the option to change the data fields.

A second configuration of navigational statistics displayed.

Distance

Receivers provide a variety of distance measurements. For receivers that calculate routes that follow roads, the distance is the distance traveled along the indicated route. For receivers that do not follow roads, the distance is the straight-line distance between two locations. Receivers can provide:

Distance to destination:

The distance from your current position to the final destination, whether along roads or in a straight line.

Changing which navigational statistics are displayed.

Distance to next waypoint in route:
The distance from your current position to the next waypoint of the route, whether the route is automatically calculated by the receiver or formed by the user using the route function.

Trip odometer:
The actual distance traveled as measured by an odometer.

Maximum and minimum elevation:
A measure of vertical, not horizontal, distance of the highest and lowest elevations reached.

Total ascent and descent:
Another vertical measurement. Total ascent is the change in vertical position while traveling up. Total descent is the change in vertical position while traveling down.

Distance to destination and next waypoint displayed.

Although some receivers can measure the vertical distance traveled, they do not have the ability to include the vertical distance in the calculation of the horizontal distance between two points, so the distance to the next waypoint is calculated as though your current position and the next waypoint were at the same altitude. When a route follows the roads of the built-in map, the distance between two waypoints inherently includes the vertical distance, because the length of the road inherently includes the change in altitude.

Distances are reported in the units you select: statute, metric or nautical. Many receivers allow the altitude to be displayed in units different from the horizontal distance. Altitude can be displayed in feet or meters. The rate of ascent or descent can be given in feet/minute, meters/minute or meters/second.

Speed

A receiver measures the time and distance between your current position and your previous position, then divides the distance by the time to get speed. Some of the speed statistics reported by receivers are listed below. The meaning of each type of speed is described to help you know what a speed statistic really means.

- Speed (SOG)
- Velocity Made Good (VMG)
- Average Speed

- Average Moving Speed
- Maximum Speed
- Vertical Speed
- Vertical Speed to Destination
- Average Ascent and Descent
- Maximum Ascent and Descent

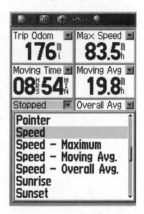

Selecting from a list to display a desired navigational statistic. Here, the statistic "Stopped" is being replaced by "Speed."

One type of speed not listed above is the "standing still" speed. Accuracy of the GPS, as discussed in Chapter 2, can make a receiver report that you are moving even when you are standing still. Receivers use averaging algorithms to eliminate false speed reports when standing still. All speeds are reported in the units you select: miles per hour, kilometers per hour or knots.

Speed, also known as **Speed Over Ground** (SOG) or ground speed, is just like the speed given by the speedometer in a car: it measures how fast you are going at that very moment. Speed does not take into consideration whether you are on course; it is a measurement of your motion regardless of direction.

Velocity Made Good (VMG) is the speed at which you are approaching your destination. VMG takes into account your present course and your destination. If you are directly on course, VMG is the same value as SOG. If you stray from course, VMG decreases and is less than SOG. The adjacent figure shows how VMG and SOG relate.

Unlike speed, VMG is not always positive. If you are moving directly away from your destination, your VMG will be the same as SOG, but it will be negative.

Average Speed is not the same as Speed Over Ground (SOG). SOG is your speed at any given moment. If one second you are moving at 25 mph (40.2 km/h), the SOG shows 25 mph. If a second later you are moving at 50 mph (80.5 km/h), the SOG displays 50 mph. Average speed divides the distance you go by the amount of time it took. Suppose you have driven your car for a while and the average speed is 25 mph. When

Relationship between SOG and VMG.

you suddenly accelerate to 50 mph, the average speed does not immediately change, but slowly starts to rise. After you have traveled 50 mph for as long as you have traveled at 25 mph, the average speed is only 37.5 mph (60.4 km/h).

Average Speed tells you how fast you really go in heavy traffic or your true hiking speed after breaks are factored in. At the start of your trip, average speed is the same as SOG. When you slow down for traffic or take a rest break, the clock keeps counting, so your average speed decreases as long as you do not move.

VMG shows you are moving in the opposite direction.

Average Moving Speed is your average speed excluding the time you stand still. Like average speed, average moving speed is calculated by dividing the distance traveled by the time it took to travel, but when your speed goes to zero, the clock that measures the time stops until you get moving again. If your travel is stop and go, the average moving speed will always be greater than the average speed.

Maximum Speed is exactly what it says: the fastest speed traveled during the trip.

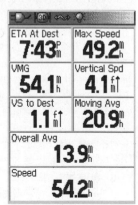

Vertical Speed is your instantaneous speed measured for up and down movements only. If you are traveling up a hill, you are moving horizontally, or forward, at some speed, and also up the hill at some other speed. The vertical speed ignores the amount you move forward and measures only the amount you change in altitude. If you stop moving or hit a flat section on the trail, the vertical speed goes to zero.

Vertical Speed to Destination is analogous to Velocity Made Good (VMG). It is a measure of your speed in the vertical direction with respect to the altitude of the destination.

A display of speed statistics.

Average Ascent and Descent is similar to the average speed above in that it is the distance of vertical movement divided by the amount of time to make the movement. It is the average rate of your change in altitude. Even though you are moving forward while moving upward, the average ascent or descent is the vertical part of your movement only.

Maximum Ascent and Descent is exactly like it sounds: the maximum rate of vertical change in position.

Direction Indicators

All directions calculated by a receiver can be expressed as a bearing. Bearings may be expressed as numbers or a cardinal direction. Numerical bearings range from 0° to 360°. The letters for cardinal directions used by receivers are N, NE, E, SE, S, SW, W and NW. The cardinal directions North, South, East and West are 0° (or 360°), 180°, 90° and 270°, respectively.

Understanding and using numerical bearings is more accurate than using cardinal letters. If the receiver reports your direction as E for east, your direction of travel may be anywhere between 67.5° (halfway between E and NE) and 112.5° (halfway between E and SE). You cannot tell if you are closer to 67.5° or 112.5°. The accuracy of the cardinal directions is limited by the number of degrees between each cardinal direction reported. A numerical bearing is accurate to a degree or possibly a fraction of a degree. When the receiver reports your direction as 90°, you are definitely going east. Bearings are described earlier in this chapter. The following definitions describe some of the direction information that you may encounter in your use of the GPS receiver.

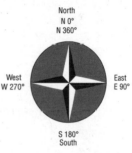

- Heading: your current direction of travel. Also referred to as track.

- Bearing: the direction from your current position to the destination. Also referred to as desired track.

- Course: the straight line from your starting point to your destination. The course does not change, even when you move off course.

- Off Course: Also known as the CrossTrack Error (XTE). This is your distance from the straight-line course.

- To Course: Manufacturer-specific. If your XTE is very small it is the same as your bearing. If XTE is large it is the direction to some indeterminate point on the course, usually the nearest point, depending on your velocity.

- Turn: The number of degrees and in what direction you must turn to arrive at the bearing.

Deviation-from-Route Indicators

Whenever your current direction of travel (heading) does not match the direction of the Goto destination or the next waypoint in an active route, you are off course. A GPS receiver can report not only how far you are off course, but what you need to do to get back on course.

Statistics like CrossTrack Error (XTE) and Course Deviation Indicator (CDI) tell you how far you have strayed from the intended course. Receivers also provide information like Course to Steer and Turn, which tell you how to get back onto the correct course. Each statistic is described in detail below.

CrossTrack Error (XTE) measures the distance between you and the direct line between the point you started from and your intended destination (course). XTE is most helpful to pilots of boats or planes or a person in an open area, because the navigator has the freedom of movement to maintain a direct course. Those on foot or in off-road vehicles usually have to go around obstacles, so they are less concerned about how far they are from the direct course and more concerned about the bearing from their present position to the destination. However, there may be times when you can maintain a straight-line course to the destination and XTE is useful. If you use your receiver on roads, XTE is meaningless because you are confined to the road. A good Map screen and routes automatically calculated to follow the roads are more useful.

The straight, solid line in the figure is the direct course between Start and Destination. The receiver tries to steer you on the straight-line course. When you stray, the XTE is the perpendicular distance between the straight line and your current position. The dashed lines up the page show the amount of CrossTrack Error to the right and left of the direct course. As you can see, XTE varies with your distance from the straight line. The point of maximum XTE is indicated for each major excursion. Whenever you are on the direct line, XTE is zero. The points of no XTE are circled.

You can always return to the direct course by traveling the opposite direction of the XTE until the XTE is zero. For example, if you have strayed to the left, you would travel to the right to return to the direct course. XTE is measured in the units you select for the receiver: kilometers, miles or nautical miles.

Regardless of how far off track you may be, the bearing reported by the receiver is the direction from your present position to the destination. If you get really far off track and it does not make sense to return to the straight-line course, simply follow the bearing that leads from your current position to the destination.

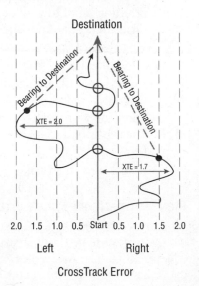

CrossTrack Error

The Course Deviation Indicator (CDI) graphically shows the amount and direction of CrossTrack Error. CDI displays are best utilized when you are required to travel a certain corridor, like a pilot of a plane or a ship. CDI is very useful for traveling such straight corridors.

The Highway Screen

CDI is most commonly displayed using the Highway screen. The Highway screen shows the straight-line course, from the starting point to the destination, as a road. The Highway screen shows your present position in relationship to the road, so you can see when you wander off course and the direction you must go to return to course.

For example, on the display you can see the Highway (thick dark line with even darker center line) leading toward Point A. Your position, the dark triangle, is 322 ft. to the right of the course, so your XTE is 322 ft. to the right of course. Furthermore, your current direction of travel is 334°, but the bearing to the destination A is 311°. If you continue your current direction of travel, your XTE will increase. The Turn statistic indicates that if you turn 23° to your left, you will again be traveling toward the destination (heading will be the same as bearing), thereby taking you from your current position to the destination at A. The Turn statistic does not tell you how to return to the straight-line course, but how much you need to turn to travel toward the destination.

CrossTrack Error in relation to bearing and heading.

If you need to maintain the straight-line course from the starting point to the destination, there is additional navigation information that can help you. The To Course, or Steer to Course, statistic provides you with the heading you need to intersect the straight-line course.

After traveling a little farther you decide you want to return to the original straight-line course. You see that your current heading is 308°. Turn indicates that the bearing from your current position to the destination is 309° (Bearing = Heading + Turn), so, to travel from your present position directly to the destination, you need to turn 1° to the right. The To Course statistic provides the direction you may go to

To Course guides to original straight-line course.

intersect the straight-line course. To Course advises you to turn 18° to the left, not to the right, so you will intersect the straight-line course at some point prior to arriving at destination A. As you approach the straight-line route (i.e., XTE decreases), the To Course gradually changes to become the bearing to the destination.

If you are close to the destination and it does not make sense to return to the straight-line course before heading to the destination, To Course will simply instruct you to travel directly to the destination.

The Course Pointer

The Compass screen may also function as a CDI. To display CDI on the Compass screen, it is necessary to enable the course pointer function, which enables the Compass screen to display an additional line of dots and a small number in the corner. The number in the corner shows the CDI limitk. The CDI limit is the maximum distance from the center line that is displayed on the screen. The CDI limit may be set to small (0.25 mi.), medium (1.25 mi.) or large (5 mi.). Should you be using metric units, the CDI limit would be 0.25 km, 1.25 km and 5 km respectively. Each of the five dots on either side of the center line represent one-fifth of the CDI limit. On the medium setting, each dot represents 0.25 mi. of deviation. The arrow continues to point the bearing, or the straight-line course from your current position to the destination.

The CDI is most useful when you have complete control over the directions of your movements and the distance to the destination is significantly greater than the CDI deviation setting. The CDI helps you stay on the straight course between two points. If you cannot stay on a straight-line course, because you have to follow roads, the CDI is not very useful. The CDI is useful in planes, watercraft or on open range (e.g., prairie, desert). Furthermore, if the distance between the two points is shorter than the CDI limit, the CDI will not really pro-vide much assistance. For example, if you are walking 2 miles (3.2 km) between two points and the CDI limit is set to 0.25 miles (0.4 km), the amount of deviation is almost the same as the distance you are walking. You can stay within the deviation limited, yet still do a lot of extra walking. However, if you are in a plane flying 1200 miles (1931 km) between two points, staying within 0.25 miles (0.4 km) of the straight-line course will save a lot of fuel.

To Course points directly to the destination.

The middle image on this page provides some valuable information about the CDI screen. In the top right corner of the Compass screen we see that the scale is 1.25 mi. As the arrow is nearly in the center of the 10 dots we know we are very close to the straight-line course from the initial starting point to the destination. Looking at the off-course display we can see that we are only 65 feet (19.8 m) off course. The arrow is pointing to 148°, which is both the Bearing and the To Course heading.

You decide to see what happens if you travel directly west. From the line on the CDI screen you see that your desired course lies just over 2 dots from your current position. The scale is 1.25 miles, so each dot is one fifth of the scale, or 0.25 miles. Two dots represents 0.5 mi., thus you know you have deviated from the straight-line course by at least that amount. You look at the Off Course statistic to confirm that you are 0.6 mi. off the straight-line course.

When you travel beyond the limits of the CDI, the line (between arrow and tail) becomes stuck on the far side of the Compass screen. You can see from the Off Course display that you are 1.71 miles off the straight-line course, which is beyond the 1.25 mile limit of the CDI. You can see that the Bearing is 146°, which is not very far from the initial 150° bearing, because of the long distance you are traveling. If you want to see where you are on the CDI, you can easily increase the CDI limit to 5 miles and then use the Compass screen to return to the original course.

As mentioned above, the CDI function displayed on the Compass screen may be very useful when you have a free range of motion and are traveling long distances.

Timers

Receivers can report a variety of times as navigation statistics.

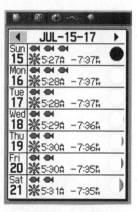

Using the compass as a CDI.

Further deviation from the straight-line course.

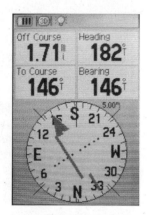

Changing the CDI limit.

Estimated Times

Most receivers provide an Estimated Time of Arrival (ETA) and an Estimated Time En Route (ETE), which is also known as Time To Go (TTG). ETA is the time of day (e.g., 10:17 am) that you will arrive at the destination. ETE or TTG tells you how much longer you must travel before arriving at the destination. ETE is measured in minutes or hours. ETA and ETE are accurate only if you are on route whether a straight-line route or a route that follows roads. Both estimated times are calculated using Velocity Made Good (VMG), which was explained earlier. Some receivers report the

Your course lies to the west.

estimated time to the next waypoint in a route along with the time to the route's final destination. Vehicle-oriented receivers report the time to the next turn in the road and the time to reach the destination.

Other Times:

Trip time, also known as **elapsed time**, measures time from the last time the timer was reset. The trip timer is used to calculate average speed, because it continues counting time regardless of whether you are moving or not moving.

Time moving is the amount of time your speed is not zero. When you stop, the time moving timer also stops. The time moving is used to calculate the average moving speed.

Time not moving is the time spent standing still. If the time on the time moving and the time not moving timers are added, they should equal the trip time timer.

Time of day. The GPS satellites keep what is known as GPS time. Receivers convert GPS time to Coordinated Universal Time (UTC). UTC is a 24-hour time measurement referenced to Greenwich, England. Greenwich is the location of zero degrees longitude. Midnight in Greenwich is zero hour on the UTC clock. The time in any area of the world is an offset from UTC. Most receivers allow the user to specify their time zone as Eastern, Central, Mountain, etc., so the receiver can convert UTC to local time. Some receivers do not ask for the local time zone, but allow the user to specify the offset between UTC and local time. For example, the offset for Arizona is –7:00 hours.

Sunrise and Sunset can also be predicted by the GPS receiver.

Glide Ratio

Glide Ratio is provided by some receivers to relate simultaneous horizontal and vertical movement. The glide ratio is the ratio of horizontal and vertical movement. Although the glide ratio is reported when you are moving on ground that changes in altitude, it is most useful if you are in some type of aircraft such as a plane or a glider.

A GPS receiver reports many different times.

Jumpmaster

Some advanced GPS receivers calculate the best position to start a parachute jump to be able to land at a desired location. The user must enter assumptions about different wind speeds at different altitudes, the jump altitude and the parachute opening altitude. The receiver estimates the route from the location where you exit the plane to a drop point on the ground. The barometer is used to measure your altitude and must be calibrated. This feature is designed for advanced jumpers who already understand the calculations the receiver performs.

Navigation Screens

The **Position Screen** is the screen where the receiver reports the coordinate of your present position. Many GPS receivers have at least one user-programmable screen that permits the user to select the information displayed. A hiker using paper maps would want the coordinate of the present position, time of day and sunset and possibly ETA to be displayed. A boater may want XTE or CDI, Course to Steer and barometric pressure shown on the screen. Most receivers have at least one page that permanently displays your position in the coordinate system you select.

Steering screens are only useful when the receiver is activated in the Goto or Route mode. It points the way to the active Goto waypoint or the next waypoint in the active route. There are three main types of steering screens.

The **Compass Navigation Screen** is also known as a pointer screen because it points in the direction you should go. It is the best screen to use when you have to go around obstacles and cannot travel directly to the destination. Hikers find it especially useful in conjunction with a compass. The arrow always points the direction from your current position to the destination waypoint or next waypoint in a route. Following the arrow leads you to where you want to go. The Compass screen can also be used to display Course Deviation information as explained previously.

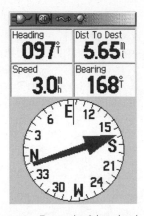

Turn to the right to head toward your destination.

The **Highway Navigation Screen** is designed for those who can go directly to their destination and who want to stay close to the straight-line course. Using the Highway navigation screen is simple because you just need to follow the road. If you travel directly to the destination, the Highway points straight up on the screen. If you stray to the right, the Highway points to the left indicating the direction you should steer to move toward the destination. The Highway screen was discussed in the Course Deviation Indicator section because it is usually combined with CrossTrack Error information. Some receivers allow the user to program the amount of CrossTrack Error allowed before an alarm sounds. Other receivers simply show the Highway on the screen and display the CrossTrack Error as a number and a letter "L" or "R." The number indicates your distance from the desired course, while the letter is your position left or right from the desired course.

The **Map Screen** provides route guidance for receivers that calculate a route that follows the roads of the built-in map. The Map screen clearly indicates the road and each turn of the route. The Map screen is discussed in detail below.

Map Screens

Built-in electronic maps make the Map screen the most useful screen of all. The Map screen indicates your position relative to all stored waypoints and all information in the electronic map. If you have street maps on your receiver, you will see what road you are on. If you have downloaded a topographic map, you will see your elevation and the topology of the surrounding terrain. Map screens offer a lot of options, such as amount of detail, auto zoom, orientation, tracks and scale.

Setting the map **Scale** permits you to zoom in and out on the map. Many receivers have buttons dedicated to zooming in and out, which makes zooming easy. Zooming is important if the receiver uses detailed maps, so look carefully at how the receiver zooms before you buy. If zooming is cumbersome or takes a lot of steps, you may want to reconsider buying that model. The level of zoom is normally displayed in the bottom corner, as a distance, such as feet or miles, or as a level of magnification.

Control over the **amount of detail** displayed on a Map screen is important because too much information can clutter the screen, but the decision as to what is too much information is a personal matter. You can adjust the amount and type of information displayed on the screen. Displaying a minimum amount of information may omit minor roads and towns. Not displaying certain types of data (e.g., railroads, local road names, tide stations) may permit other information to be displayed without cluttering the screen.

Denver shown at a scale of 20 mi.

Many receivers provide an auto display mode, which lets the receiver determine when a feature such as a waypoint or a minor road appears on the screen. The auto mode is governed by the overall amount of detail you specify, whether it be more or less. The auto mode adjusts the amount of information to the amount of area displayed on the screen. As you zoom in, more detailed information appears, and disappears as you move out. This level can be manually set by the user for waypoints, POIs, street labels and land cover. You can also set the text size for these same points.

Most amount of detail, shown at a scale of 3 mi.

Auto Zoom enables the receiver to automatically zoom in as your speed decreases, to provide more details on the map, or to zoom out as your speed increases to show a larger area. The auto zoom mode is very useful.

Orientation is an important aspect of map navigation. The map can be oriented so that north, your direction of travel (track) or your destination is always at the top of the screen. "North up" orientation makes it easier when using a paper map. "Track up" orientation always puts the direction of travel at the top of the screen, so that as you turn, the map turns to maintain the upward direction. Track up provides a quick reference to locations around you. If a point is at the top of the screen, it lies ahead of you; if at the bottom, it is behind, etc. Track up is useful when using a receiver

Least amount of detail.

in a vehicle. Course up orientation means that the destination is always positioned at the top of the screen. If you have complete freedom of movement and can stick to a straight-line course, the course up mode shows your deviations from course as discussed above in the sections on XTE and CDI.

Latitude/Longitude grid may be displayed on the Map screen of some receivers. This function may not be as useful as displaying your current coordinate in the coordinate system of your choice.

Tracks. Most receivers periodically store your location in what is known as a track log, or track. If the receiver is continuously on during your journey, the track represents your progress along the way. Receivers can be set up to record waypoints in the track log at a set time interval or a set distance interval, or to automatically decide the best method for storing waypoints in the track log, which may be a combination of time or distance, depending on your speed and movement. Each entry in the track log contains the waypoint of the location where the entry was recorded and the time it was recorded. It may also contain the elapsed time and distance between track points.

Tracks may be transferred from a receiver to a computer. The computer can display your track on a map. Track logs may contain hundreds or tens of thousands of track points. Many receivers allow the track to be reversed and used to navigate back over the path traveled. Tracks may also be displayed on the receiver's screen or not displayed if they clutter the screen.

Tracks are not important if you do not leave your receiver on for the duration of the trip, but if you do leave it on, tracks can accurately show you where you have been and how long it took. Some receivers allow tracks to be saved, much like a route, for later navigation. Setting the receiver to store a track and later

Changing from North Up to Track Up.

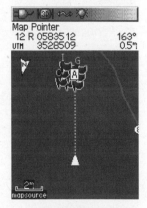

Heading toward the south with the destination at the top of the screen.

Tracks set to record every 30 seconds.

converting it to a route is a very convenient way to form a route of your trip.

Although many receivers provide a lot of memory for track logs, the number of points that can be stored is still finite. If the track log fills up, it can be set to either stop saving new track points or to overwrite the track points stored in memory starting with the points at the beginning of the track. If you want to save a track or convert it into a route, be sure you do not overrun the size of the track log, or you will not have a complete record of your trip.

A track log is very useful in search and rescue operations. Each search team should keep a receiver on and continuously record a track log. When the team returns, the track can be transferred to a computer as a permanent record of where that team has searched.

Saved tracks can distinguish different legs of a trip.

Customizable Screens

As mentioned above, many receivers allow the user to select the navigation statistics displayed on a screen. The boxes that display the information are called data fields. Customization allows you to see on a single screen all the information that is important to you. Your mode of travel determines what information will be important, so customizing the data fields enables you to use various modes of transportation, whether it be on foot or in a vehicle, and still see all the important information you need without paging through screens.

Choosing the number of data fields to be displayed.

Miscellaneous Functions

Initialization

When the receiver is turned on for the first time or after it has been off for several months, it needs to know its approximate location to be able to lock onto the satellites. Most receivers make initialization simple by allowing you to select your approximate position from a list of countries in the world. Some receivers allow you to select your approximate position from the built-in map.

When you first start up a receiver, it may take up to 12.5 minutes before it locks onto the satellites and provides a position. The satellites broadcast an almanac that contains the current position of all the satellites. The receiver maintains a record of the almanac in its memory. When a receiver first starts up, it determines whether its almanac is up to date. If it is current, the receiver knows where the satellites are and can immediately start searching for available satellites. If its almanac is old, the receiver must have an updated almanac before it can do anything. Downloading a new almanac takes a long time. The time it takes from the moment you turn the receiver on to the time it calculates its position is called **Time To First Fix** (TTFF).

Displaying pertinent information on each screen can save you switching back and forth between different screens.

One company provides a special almanac (also known as ephemeris) that decreases TTFF to 10 seconds. The almanac broadcast from the GPS satellites predicts future satellite positions accurately for about four hours after download. The special almanac accurately predicts future satellite positions for up to seven days after download. Transferring the special almanac to the receiver before using it helps the receiver find the satellites faster, so a position can be calculated with less delay. This capability does not appear to be currently available in any GPS receiver, but may be in the future.

Man-overboard (MOB)

The Man-overboard function is useful if you have to quickly mark a spot and return to it. Imagine you are on your boat, traveling across a lake at night, when suddenly you hear a cry that indicates someone just fell overboard. Being a quick thinker, you press the MOB button. The receiver immediately records your present position, then it instantly switches to the navigation screen to direct you back to the location recorded. By the time you slow down and get the craft turned around, the receiver is already pointing the direction back to the location. When you arrive at the MOB waypoint location, you may begin searching for the lost passenger.

The usefulness of the MOB function is not limited to boats. Imagine you are part of a search and rescue team. As you do a low-altitude flyby in a fixed-wing plane you see what looks like the lost hikers you are looking for. You press the MOB, which marks their position and gives you the coordinate to call in to the ground teams.

Calculations

Solar and Lunar Calculations

Many receivers calculate the times for sunrise, sunset, moonrise, moonset and moon phase based on the date and location. If you are planning an expedition months in advance or to an unfamiliar area, it is nice to know when it will be light or dark or if the moon will provide any light at night. Some receivers also allow you to sequence (press the play button) through the days of a month while displaying the changes in moon phase, sunrise, and sunset.

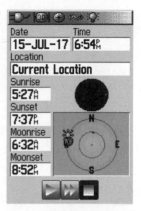

No moon will shine for the selected date and position.

An entire week in one screen.

Hunting and Fishing

Some receivers can predict the best times for hunting and fishing for a particular date at a particular location. Unless specifically stated otherwise, these predictions for good hunting and fishing times correspond to the amount of light available and not to any information about animal behavior at the location. A GPS receiver can accurately calculate the rising and setting of the sun and moon and the phases of the moon. However, it knows nothing about cloud cover, temperature or ecological conditions. Unless the manufacturer specifically provides additional information, the receiver cannot know about the feeding habits of fish or the migration habits of elk or deer. The receiver's predictions about the best times for hunting and fishing are a statement about the best time for you to be in the field and not if you will meet any animals.

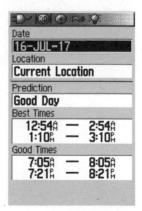

A prediction of a good day for hunting and fishing.

Tides

If you boat or paddle on the ocean, you will want a receiver capable of calculating the tides. Ocean tides are caused primarily by the pull of the moon's gravitational force on the earth. GPS receivers can accurately calculate the moon's movements, so it can also predict the tide.

Coordinate by Reference or Projection: "Sight and Go"

The coordinate of a waypoint can be calculated by specifying the bearing and distance from any other waypoint. Suppose you are on the safety patrol at a ski resort when an exhausted skier arrives at the lodge. There has been an accident and all the skier can tell you is that he came from "that" direction and skied for about an hour to get to the lodge. You turn on the electronic compass in your receiver to get a bearing and make a rough estimate of the speed and distance traveled. You create a new waypoint for the accident site that is the distance you estimated at the bearing you measured from your present position. The receiver calculates the location's waypoint based on that information. Now you can use the calculated coordinate to easily find the location on a map and start a rescue operation close to the probable accident site.

"Sight and Go" function permits the user to set a compass heading for a course or project a waypoint.

Area Calculations

Some receivers can calculate the area between several waypoints or inside the track lines made during your trip. The points used to define the area must form a closed area for the calculation to be done. Minerals prospectors can use the area calculation to mark the corners of a claim and measure the area.

Projecting a waypoint at a bearing of 115° and a distance of 12.6 mi. (20.34 km) from your current position.

Time

Universal Time Coordinated (UTC) is measured by precise atomic clocks. Ideally, UTC should match earth's solar time as measured by the earth's rotation and orbit around the sun. Unfortunately, the solar day has lengthened by 1.7 ms every century. The increase is not noticeable to humans, but it is enormous to an atomic clock. The UTC clocks are occasionally slowed by 1 second (e.g., a leap second) to keep UTC close to solar time. GPS clocks keep GPS time, as mentioned above, which does not track earth time, so GPS time does not add leap seconds. The result is that GPS time is not the same as UTC. The GPS clocks started at zero hour on January 6, 1980. Since that time, 14 leap seconds have been added to UTC, so GPS time is 14 seconds ahead of UTC. Again, 14 seconds does not seem like much, but when making celestial observations, use UTC and not GPS time.

Move to highlight the letter "C" and select it.

Data Entry

Hand-held GPS receivers are designed to be as small as possible, so they do not have a keyboard like a computer, a PDA or a cell phone. Unless you do all of your data entry on a computer and then transfer it to the receiver, you have to type in one number or letter at a time. Data entry is straightforward and fairly fast because receivers either serially cycle through the alphabet or provide a table of the alphabet that reduces the number of keystrokes required for entering data.

Blackberries and other PDAs have accustomed us to QWERTY keyboard layouts in hand-held devices. Unfortunately, the alphabet tables displayed on receivers are not QWERTY, so prepare yourself for a bit of confusion when first using an alphabet table on a GPS receiver.

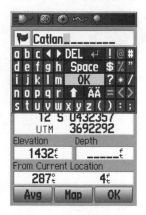

Press OK to accept the waypoint name.

5 Using UTM Coordinates on a Hiking Trip

There are two ways to use a GPS receiver for hiking. You can immediately start off on your adventure and mark waypoints along the way to use for the return trip or you can plan the trip in advance and let the receiver guide you along a preplanned route. Most people use the first method because it is simple and does not require any planning. Using the second method allows you familiarize yourself with the area before you arrive and helps you prepare for any known challenges that exist. A computer can make the planning easier by marking waypoints on the computer and transferring them to the receiver for the trip.

This chapter introduces the UTM grid and how to manually form a route on the receiver. The hiking example shows how to blend GPS navigation with traditional navigation skills.

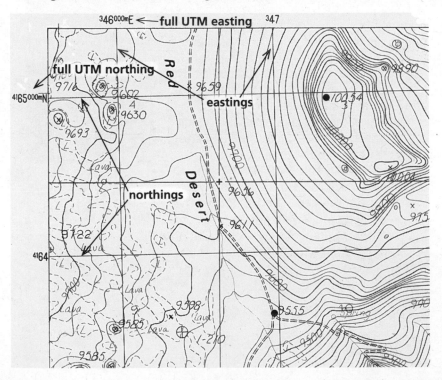

Introduction to the UTM Grid

Enough information on UTM coordinates is given in this chapter for you to determine the coordinate for any point on a map and to use the coordinates with a GPS receiver. Additional information on the UTM grid is given in Chapter 6. The latitude/longitude grid is explained in Chapter 7.

All maps have a grid that uniquely defines every point on the map. The topographical map on page 70 shows the northwest corner of Henrie Knolls quadrangle, Utah (7.5-minute series, scale 1:24,000). It is printed with the UTM grid and has tick marks for the latitude/longitude and state grid.

The lines indicated with arrows form the UTM grid. The numbers along the top of the map are called eastings, which provide an east/west position. The numbers on the left side of the map are called northings, which provide a north/south position. Below is a quick introduction to eastings, northings and UTM coordinates.

Eastings

- Increasing easting numbers means you are moving east.
- Full easting coordinate number: 346000m.E.
- Distance between 346000m.E. and 347000m.E. is 1000 m (1 km).
- The large numbers are an abbreviation. On this map:
  46 means 346000m.E.
  47 means 347000m.E.
- The last three numbers stand for meters.
  Distance between 347180m.E. and 347721m.E. is 541 m.
  Distance between 347180m.E. and 352721m.E. is 5.541 km.

Northings

- Increasing northing numbers means you are moving north.
- Full northing coordinate number: 4165000m.N.
- Distance between 4164000m.N. and 4165000m.N. is 1000 m (1 km).
- The large numbers are an abbreviation. On this map:
  64 means 4164000m.N.
  65 means 4165000m.N.
- The last three numbers stand for meters.
  Distance between 4164300m.N.and 4164560m.N. is 260 m.
  Distance between 4164300m.N.and 4202560m.N. is 38.260 km.

The UTM grid is based on meters, and grid lines are always 1 km (0.62 mi.) apart on large scale maps. Estimating distance on a map is easy with the UTM grid because the distance between grid lines is the same distance on the ground. As you will see in Chapter 7, the distance between latitude/longitude grid lines does not directly correlate to a distance on the ground.

UTM Coordinates

- The form of a UTM coordinate is zone, easting and northing.
- The zone is printed on the map. For this map it is 12.
- For some receivers, the zone is "12 S" or "12 N." See Chapter 6.
- The complete UTM coordinate of the hill 10,054 ft. is
  12 347400m.E. 4164900m.N., or, including the zone letter,
  12 S 347400m.E. 4164900m.N.
  Abbreviated: 474 E. 649 N.
- The complete UTM coordinate of the junction between two unimproved roads with the elevation of 9555 ft. is
  12 S 346900m.E. 4163600m.N.
  Abbreviated: 469 E. 636 N.

For an explanation of the zone letter "S," see page 88.

Guidebooks that utilize maps with a UTM grid sometimes identify unnamed features by their abbreviated UTM coordinate. Accordingly, the intersection of the roads identified above would be referred to as Grid Reference (GR) 469636 and the hill as GR 474649.

Accuracy of UTM Grids

Some recreational users question the accuracy of UTM grids. The UTM grid is a projection of the earth's curved surface onto flat sheets of paper, and as a result there are some inaccuracies across each of the 60 zones (see the explanation of zones in Chapter 6). However, the error is so small that it is of no concern to users of civilian hand-held receivers. The inaccuracies of GPS receivers due to ionospheric interference and satellite geometry described in Chapter 2 are greater than the inaccuracies of the UTM grid.

If you really need to correct for inaccuracies in the grid, some receivers allow you to enter a value for grid declination, which is the difference between grid north and true north. Some maps, including many of the USGS 7.5-minute series, have the angular difference between grid north and true north printed on the map (see figure on pg. 49).

UTM and a GPS Receiver

You now know enough about the UTM grid to read coordinates from a map, but there are a few additional points to know about using UTM coordinates with a receiver.

- GPS receivers need the complete coordinate numbers. A receiver does not understand abbreviations, because they are too map-specific. For example, the coordinate for Water Tower is entered as 12 S 0405676E.3689401N.

- When entering the coordinates, you do not enter the "m·E." or "m·N." The receiver automatically displays them.

- Some receivers require seven digits for both easting and northing values, while other receivers only require six digits for the easting. The first digit in a seven-digit easting will always be zero.

- For ease of use, when taking coordinates from a paper map without using a ruler, round your numbers off to the nearest 10 or 100 m value. On a larger scale map, it is easy to round to 10 m, while a smaller scale map may make rounding to 100 m easier. For example, if the exact coordinate of a landmark, as measured by a ruler, is:

12 S 506913m.E. 4615672m.N.,

it may be more easily read from a map, without a ruler, as:

12 S 506910 4615670 or
12 S 506900 4615700,

depending on the scale.

Water Tower Waypoint.

Baltimore Waypoint.

When measuring coordinates off a paper map using a ruler, be as accurate as the ruler allows. Rulers for smaller-scale maps (1:24k, 1:25k) generally have 2 m resolution. Rulers for large-scale maps (1:63.36k, 1:50k) have marks every 50 or 100 meters. Electronic maps provide coordinates in 1 m intervals.

A Hiking Trip in the Mountains

Navigation Plan

The route used in this example could be navigated using compass alone because of the identifiable landmarks such as mountains and streams. The route could also be navigated using a receiver by marking waypoints along the way to guide you on the return trip. This example illustrates how to preplan a trip using a map and how a GPS receiver can complement your present navigation skills. The scenario shows when and why you might rely on the receiver alone, and when it may be appropriate to use a compass and altimeter.

The example also demonstrates that all aspects of a trip cannot be planned in advance. At one point you have to cross a stream, but you have never been there before and do not know the best place to cross, so your navigation plan allows you to search for a crossing place and still stay on course.

The trip's waypoints are marked on the map on page 75. You will arrive by helicopter at the lake near point #1 in late June. You want to set up camp before the sun goes down, so you use the receiver to calculate sunrise and sunset. The receiver reports that on that day, the sun rises at 4:29 am and sets at 9:28 pm. You arrange the helicopter to drop you off two hours before sunset and plan to catch fish in the lake the next morning.

Your ultimate destination is point #8, but you plan to take two days to get there because you will photograph wildflowers along the way and want time to search for the rare Alpine poppy that grows on Calumet Ridge. On the first day, you will hike from point #1 and camp overnight at #6. The next day, you will continue to point #8, where you will join your colleagues for four weeks of botanical studies at a well-established camp. The bush in the area is dense spruce, so you plan to stay above the tree line as much as possible.

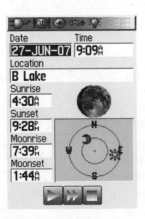

B Lake sunrise/sunset.

Your navigation plan combines your GPS receiver with a compass and altimeter to conserve the batteries because you will be gone a month and are taking only one fresh set for the return trip. If you have a receiver that has an electronic compass or altimeter, be sure to turn off the GPS receiver portion when using only the compass and altimeter, to conserve energy. It would be wise to carry a separate compass and altimeter to navigate if your receiver malfunctions.

EDITION 2 83 E/3

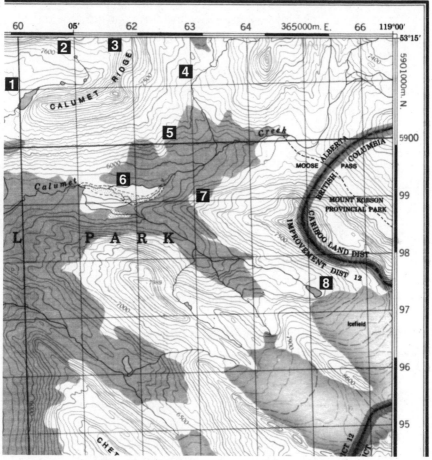

Northeast corner of Mount Robson, British Columbia, Canada (1:50,000 scale).

- **Travel from point #1 to point #2:**
  Use the GPS receiver the entire way to travel directly to the lake.

  The receiver will keep you on the right course, thereby saving energy and time.

- **Travel from point #2 to point #3:**
  Use the GPS receiver the entire way to travel directly to the top of the ridge.

- **Travel from point #3 to point #4:**
  Use the compass and altimeter.

  You do not need to get exactly to #4.

You will be descending steep scree with intermittent cliff bands. In a mist, this will be tricky and time consuming.

Use the compass to walk toward #4 until you reach an altitude of 6800 ft.

- **Travel from point #4 to point #5:**
Use the altimeter, with the GPS receiver occasionally.

You want to travel almost in a straight line from #4 to #6.

Point #5 is situated just before you enter the trees. You want to arrive as close to #5 as possible. Use the altimeter to drop gradually to an altitude of 6400 ft. as you approach the trees. Once you are close to the trees, check your position with the receiver. If necessary, use the receiver's Goto function to guide you the last part of the leg to #5.

- **Travel from point #5 to point #6:**
Use the compass.

You have to walk through trees. You have a newer model receiver and believe the receiver will pick up the satellite signals through the trees, but you plan to rely on your compass just in case it does not.

Walk the bearing between #5 and #6.

When you reach the clearing at the creek, use the receiver to verify your position.

- **Travel from point #6 to point #7:**
Use the compass and altimeter.

You are not sure where you can cross Calumet Creek, as the runoff is high. Follow it upstream until it can be crossed.

Cross the creek and try to get a new location reading using the receiver.

If the receiver locks onto the satellites, walk the bearing from your present position to the tree line at #7. If the trees block the satellite signals, travel southeast (true north reference, not magnetic) to the tree line.

- **Travel from point #7 to point #8:**
Use the compass and receiver.

Follow the contour of the tree line, and then the stream when you reach it, until you reach the upper valley where the terrain levels off. Use the receiver to get an occasional fix, but for the most part, follow the stream to the lake.

In good weather the entire route could be navigated by sight, but in thick cloud or driving rain the receiver is very useful.

Entering Waypoints

The plan looks feasible, so it is time to enter the waypoints into the receiver. Even though the receiver is not used to guide you to every waypoint, there are several good reasons to enter all of them into the receiver. The first is because the receiver automatically calculates the bearing between each point. Of course, you could measure the bearings from the map, but it is so much easier with the receiver, especially when it automatically compensates for declination. Second, the receiver can also calculate the distance between each waypoint. Only the route between #7 and #8 is not direct, so the sum of the distance between each waypoint will be close to the actual distance traveled. Because the receiver does not include changes of altitude in its distance calculation, it is impossible to get a complete picture of the hike's difficulty using a paper map, but the total distance does provide an indication. If you use a computer to select the waypoints, most map programs will show a profile of the hike to provide an idea of the hike's difficulty. Third, receivers show your present position relative to any stored waypoints, so you will be able to see where you are in relation to your planned route. A final reason to store each waypoint is to prepare for the possibility of bad weather, which would mandate more reliance on the receiver than anticipated.

Waypoints entered and ready for use in the field.

Use the map to estimate the UTM coordinates for each point. When estimating, remember that 100 meters is one tenth of the UTM square grid shown on the map. The waypoints are represented by the dot next to the numbers on the map. The UTM coordinate for each point, rounded to the nearest 100 m, is given below along with a name:

Point	Zone	Easting	Northing	Name
#1	11 U	360100m.E.	5900800m.N.	B Lake
#2	11 U	361000m.E.	5901600m.N.	S Lake
#3	11 U	361900m.E.	5901800m.N.	C Ridge
#4	11 U	362700m.E.	5901100m.N.	Flat
#5	11 U	362400m.E.	5900000m.N.	Woods
#6	11 U	362000m.E.	5899300m.N.	Stream
#7	11 U	363000m.E.	5898800m.N.	O Ridge
#8	11 U	365200m.E.	5897500m.N.	Camp

77

Before you enter the data, initialize the receiver to the following settings:

- **Map Datum: North American Datum 1927 (NAD. 27)**
  If your receiver splits NAD 27 into separate settings for Alaska, Canada, Central America, etc., select **NAD 27–Canada**.

- **Units: Metric**
  The UTM grid is based on the meter, so it is much easier to use the map if the distance is also set to metric units.

- **Coordinate Grid: UTM**

- **North Setting: Magnetic North**
  A compass is used to get between several points. If the receiver is set to report direction as magnetic bearings, those can be dialed directly into the compass without compensation for the declination.

- **CDI Limit: Small**
  If the Course Deviation Indicator's tolerance is selectable, set it somewhere between 250 and 500 m. The receiver's Goto function will be used to navigate to points #2 and #3. A CDI limit of 250 to 500 m means that at most you can stray 250 or 500 m from course before the receiver warns you.

- **WAAS Setting: Off**
  WAAS does not provide coverage in this area, so turn WAAS off so as not to introduce error.

Enter each coordinate into the receiver or transfer them from the computer. Form the waypoints into a route so the receiver will calculate the distance and bearing between each waypoint. Each receiver displays routes differently, but the information provided is shown below. The desired track is the bearing to the next waypoint. The distance is expressed in kilometers because the receiver's units were set to metric.

Name	Point	Desired Bearing	Distance km
B Lake	#1		
		28°	1.2
S Lake	#2		
		56°	0.9
C Ridge	#3		
		110°	1.1
Flat	#4		
		174°	1.1
Woods	#5		
		188°	0.8
Stream	#6		
		95°	1.1
O Ridge	#7		
		99°	2.6
Camp	#8		

Each segment in a route is called a leg. The distance of the first five legs of the trip, from B Lake to Stream, is 5.1 km (3.2 mi.), while the total distance, if hiked in a straight line between each point, is 8.8 km (5.5 mi.). The receiver is now ready for the trip and you are ready to test it in the field.

Waypoints forming the route.

In The Field

The day finally arrives. After a spectacular flight, the helicopter lands near the lake and you disembark with plenty of time to set up camp. Just as planned, you are fishing in the lake the next morning at 4:00 and catch a fine breakfast. When it is time to go, you turn the receiver on, but it seems to take much longer than usual to lock onto the satellites. A receiver can take up to 12.5 minutes to get a position fix in three circumstances. If a receiver has been moved more than 300 miles from the location where it last locked, or it has not been used for a few months, or it lost its memory (e.g., battery failure), it must download the almanac before it can calculate its position. This Time To First Fix (TTFF) was described in Chapter 4. The helicopter transported you more than 300 miles from the last position fix, so the receiver is taking extra time to lock.

The next thing you notice is that the altitude is not even close to the value stated on the map. You suspect the satellite geometry is less than optimal, but you brought a separate electronic altimeter, so you are prepared. From the map, you determine your altitude is about 7200 ft. and you calibrate your altimeter accordingly.

It is time to get going, so you activate the Goto function and tell the receiver to guide you to point #2, which is named S Lake. As you walk along, the receiver reports that the desired track (bearing), as you already knew from the route information above, is 28° and your present track (heading) is also 28°, so you are on course. Also, the arrow of the Compass navigation screen is pointing straight ahead, which confirms that you are on track.

Compass screen to S Lake.

After a while, you start to look around for wildflowers and do not pay much attention to the receiver. Once you finally look at it, you notice your direction (heading) is now 68° and the arrow on the Compass screen points to the left. You need to need to turn to the left to move directly toward S Lake. You also notice that the bearing to get to the lake has changed from 28° to 24°.

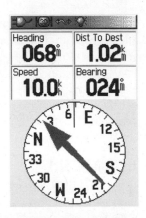

Off course already.

Before you correct, you switch to the Highway screen and continue walking just to see what happens. After walking a short distance, the Highway screen shows you are off course by 95 m. The triangle indicates that your present position lies 95 m to the right of the line going up the Highway. The darker area represents the path leading to S Lake. You also notice you have wandered so far afield that the bearing to S Lake is now 22° and not 28°. The Turn statistic informs you that you must turn 45° to the left to travel directly to S Lake.

Using the Highway screen.

You switch to the Map screen. It shows both B Lake and S Lake with a straight line representing the desired direct route between them. Your position is marked by the triangle which lies to the right of the desired route. Being off course slightly is not really a problem. You simply turn your direction of travel to the left to a bearing of 22° (as shown on the previous Compass screen) and keep going. As you turn to change course, the triangle turns to show that you are headed directly toward S Lake. You note that your estimated time en route (ETE) is about 5 minutes, so it will not be long before you arrive.

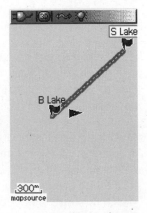

Your position on the map.

But then you start to wander again. There are more flowers with greater variety than you expected. With a few stops here and a quick picture over there, you soon find you are walking a bearing of 352° and the arrow on the Compass screen points to the right, which means you need to follow the arrow to the right to travel toward S Lake. So much for going directly to the lake as you had planned. There is just too much to see and do.

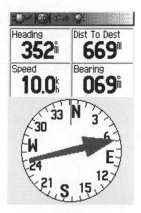

You switch to the Map screen. Your wanderings took you off course to the left. The bearing to S Lake is now 62°. You will have to turn to the right to get to S Lake, but you just spotted a large patch of wildflowers farther to the left and you must go see them. You watch the screen as you walk to the flowers. You are getting farther and farther off course to the left of the original straight line between B Lake and S Lake. By the time you reach the flowers you are 430 m from course, but the flowers are beautiful and after you take pictures you switch to the Highway screen, determined to go straight from where you are now to S Lake.

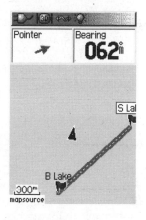

You turn to the right until the turn statistic shows 0°. Walking a track of 72° shows the S Lake waypoint straight ahead on the Highway screen. You can still see the dark road with its center line because you are not so far off course that it is off the screen.

You stay on the bearing leading to S Lake and soon the receiver beeps and announces that you are getting close to the destination. You can see the lake to the right and it is approximately 100 m away, just like you planned when you picked the coordinate for the S Lake waypoint. You continue on the course until the screen shows your position directly at S Lake.

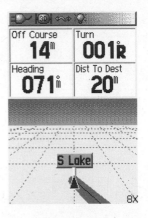

You plan to use the Goto function to get from S Lake to C Ridge as the destination. The Compass screen shows that you must bear to the right to go straight to C Ridge.

You switch to the Map screen, where you can see the B Lake, C Ridge, Flat, Woods, Stream and O Ridge waypoints along with your current position at S Lake as a triangle. Out of curiosity you move the map cursor, the white arrow, so it touches the O Ridge waypoint. The Map screen adds a new line of information that tells you that O Ridge is 3.44 km away at a bearing of 123°. It also gives the UTM coordinate of where the cursor is pointing, which in this case is the O Ridge waypoint. You move the cursor to several other locations on the map and see their distances, bearings and coordinates. If a downloadable electronic map had been available for the area, it would have been easy to plan the trip and get the waypoint coordinates using the receiver instead of using a paper map, but unfortunately, electronic maps are not available for some of the more beautiful places.

You now have a much better understanding of how to do basic navigation using your GPS receiver. First, you get the coordinate of the destination you want to visit. The coordinate may come from a paper map, an electronic map or a website (e.g., geocaching). Next, you enter the coordinate into the receiver. Entering may be done by manually typing the coordinate into the receiver or by transferring it from a computer. Then you activate the Goto function to lead you to the destination. The receiver guides you to the destination, either along a straight line or following roads, from your present position to the destination. If you do not follow the course indicated by the receiver, the receiver tells you how far off course you are and what to do to get back on course.

You were walking a bearing of 75° when you arrived at S Lake, but the bearing to C Ridge from S Lake is 56°, so the Highway in the navigation screen points to the right, showing that you need to change your course and hike off to the right.

On this leg, you pay a lot more attention to the receiver and you are able to walk the bearing fairly closely. You are able to stay on course until you get about halfway to C Ridge, where you encounter a very boggy section of ground. Your foot sinks so deep into the mud that you doubt you will be able to make it to C Ridge without getting stuck, or worse, falling down. You scan the horizon and notice solid ground far to the left that may provide passage.

You decide to leave your receiver on just to see what the navigation screen says as you take the long detour to get around the mud. To get to the solid ground, you travel at an angle of 90° off the straight line between S Lake and C Ridge, which makes your bearing 326°. The arrow in the Compass navigation screen signals a hard right, but there is nothing you can do to follow the course indicated by the receiver as long as you are walking around the mud. You switch to the Highway screen to discover that you are so far off course that the middle of the Highway does not even appear on the screen; however, the turn statistic shows that you must turn 117° to the right to return to course and the bearing from your current course to C Ridge is 83°.

At long last you reach rocky ground and you switch to the Map screen to see that you really went a long way off course to get past the mud, but the beauty of the GPS receiver is that you know exactly where you are and the bearing to C Ridge is now 102°. Taking the detour with a compass alone and still getting to C Ridge would be much more difficult. The rest of the way is fairly flat and it all looks like solid rock, so you should be able to go from where you are directly to the destination. You look at the Highway navigation screen. You are off course to the left by 600 m. You can see the Highway with its center line and you can also see that you are now headed directly for C Ridge.

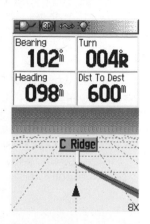

The Compass and Highway steering screens each play their part. The Highway screen shows how close you are to the straight line between two points, and the Compass screen always tells you which direction to go regardless of your proximity to the direct course. The Highway screen is best used in conditions where you have the freedom of movement to always maintain the direct course to the destination, like on water or in the air. The Compass screen works well in situations where you have to go around obstacles that lie in the straight line path.

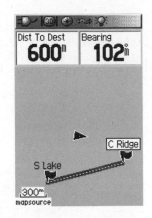

Paying attention, you continue from your present position and quickly arrive at point C Ridge. A look at the Map screen also shows the detour you had to take.

The plan to get to Flat was to walk in the general direction until you reach an altitude of 6800 ft. The bearing between C Ridge and Flat is 110° with reference to the magnetic pole, so you turn the compass housing until the number 110 lines up with the direction arrow. You take a sighting and start to walk. Your altimeter reads 8180 ft., which corresponds closely to the map. Point C Ridge does not lie directly on any of the map's altitude lines, but you know it lies somewhere between 8100 ft. and 8200 ft. The altimeter updates its electronic readout only once every two minutes, but the descent is not steep and the readout is generally up to date.

When the altimeter reads 6900 ft., you take a two-minute break, and when the altimeter updates you discover your altitude is 6800 ft. The receiver tells you that you are within 100 m of Flat. With your compass, you set a bearing of 174° and resume hiking, keeping an eye on the altimeter. At 6400 ft. you are close to the trees and the Map screen shows your position as very close to the waypoint Woods, point #5 on the map.

Next you sight a bearing of 188° and dive into the bush on the way to Stream. The plant life is a lot thicker than you thought it would be and after a while you begin to wonder if it would have been faster to have gone around the trees and followed the stream up to the Stream waypoint. You get out the receiver, but it will not lock because the foliage is so dense. There are no clearings where you might get a GPS fix, so you continue as close as possible on the bearing. Soon you are out of the trees, your receiver locks and the Goto function leads you to the Stream waypoint, #6 on the map, where you spend the night.

In the morning you execute your plan by following the stream until you find a good place to cross. Once again the receiver will not lock in the bush, so you follow the contingency plan of walking southeast to get out of the trees, but you need to set your compass to get a sighting. You know the declination is east 22° and that southeast in relation to true north is 135°. You repeat the phrase "East is least, West is best" to remind yourself that to convert from true north bearings to magnetic, you subtract east declinations and add west declinations. The declination is east, so you subtract 22° from 135° to get a southeast magnetic bearing of 113° for the area. You set your compass and do your best to get through the bush as fast as possible. Once you are through, you follow the tree line and then the stream directly to the lake without using either compass, altimeter or receiver. During the next month at the camp, you use your receiver to record the locations of wildflowers and to explore the icefields to the south and east.

When it is time to meet the helicopter, you use the map to plan a new route back to the lake that avoids traveling through the bush.

6 More UTM, and Navigating in a Whiteout

The beauty of the Universal Transverse Mercator (UTM) grid is its ease of use. It is simple to read eastings and northings directly from a paper map without the aid of a ruler. However, there are elements of the UTM grid that need more explanation, such as the origin of the zone number and letter used in the previous chapter.

The UTM grid splits the world into 60 zones that are each 6° wide. Zone 1 starts at west longitude 180°, which is the same as east longitude 180° as shown in the figure below. The zone number increases by 1 for every 6° interval until the entire circumference of the globe is covered. The last zone is number 60. The area of each zone is figuratively peeled off the globe and flattened to make a two-dimensional map. During the flattening process, the zone loses its relationship to a sphere, so UTM coordinates are called false coordinates. The latitude/longitude grid, unlike the UTM grid, relates to a sphere, so it is called a geographic coordinate. The transverse mercator projection provides a uniform grid for the entire earth; however, UTM maps do not cover the areas around the North Pole, above north latitude 84°, and the South Pole, below south latitude

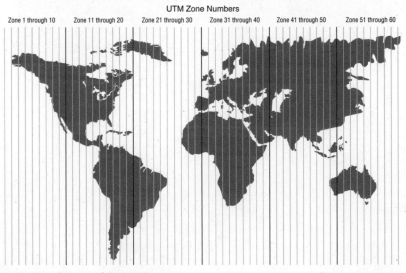

UTM Zone Numbers

Zone 1 through 10   Zone 11 through 20   Zone 21 through 30   Zone 31 through 40   Zone 41 through 50   Zone 51 through 60

Approximate locations of the 60 UTM zones.

80°, because maps of the poles are drawn with the Universal Polar Stereographic (UPS) grid. The UPS grid is discussed in Chapter 13.

Each UTM zone has a horizontal and a vertical reference line. UTM easting coordinates are measured from a vertical line down the middle of the zone called the zone meridian. Each 6° zone is split directly in two by the zone meridian. Zone 1, as shown in the figure on the right, is bounded by west longitude 180° on the left and west longitude 174° on the right. The middle of the zone lies on the W 177° longitude line, which is 3° toward the center from each side. Zone 2 is bounded by W 174° and W 168°, with its zone meridian located at W 171°, and so forth for each zone.

The meridian of every zone is always labeled 500000m.E. An easting greater than 500000m.E. lies east (to the right) of the meridian, while anything less than that

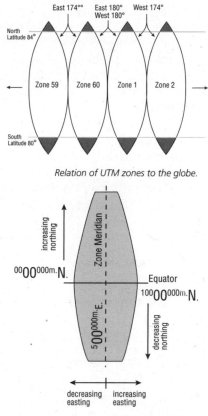

Relation of UTM zones to the globe.

UTM zone meridians.

lies to the west (to the left). The value of an easting coordinate is its distance from the zone meridian in meters. The easting 501560m.E. is 1560 m east of the meridian because it is 1560 greater than 500000m.E., whereas the easting 485500m.E. is 500,000 – 485,500 = 14,500 m west of the meridian. A valid easting for a given zone will not be less than 166640m.E. nor greater than 833360m.E. Easting coordinates always increase as you move east and decrease as you move west.

Northing coordinates are always measured relative to the equator, which is the horizontal reference line in each zone. The northing value assigned to the equator is 0000000m.N. for locations north of the equator and 10000000m.N. for locations south of the equator. You have to know if the location of interest lies above or below the equator. How some GPS units indicate whether a northing coordinate lies above or below the equator is described below. A northing coordinate for a location north of the equator is simply its distance above the equator. A northing value of 5897000m.N. indicates a location that lies 5,897,000 m north of the equator. A valid northing for a position above the equator will lie between 0000000m.N. and 9334080m.N.

The northing coordinate for a location south of the equator indicates the location's distance below the equator, but the equator is assigned the northing value of 10000000m.N. The northing value of 5897000m.N. indicates a location that lies 10,000,000–5,897,000 = 4,103,000 m south of the equator. Valid northing coordinates for the southern hemisphere lie between 1110400m.N., at the very southern end of a zone and 10000000m.N. at the equator. Regardless of whether you are above or below the equator, northing values increase as you move north and decrease as you move south.

GPS receivers use three different ways to show the hemisphere of a UTM coordinate. All the coordinates below describe the exact same place in zone 11:

11	360100m.E.	5900800m.N.
11 N	360100m.E.	5900800m.N.
11 U	360100m.E.	5900800m.N.

The first coordinate does not visually tell the user the hemisphere. When the coordinate was entered into the receiver, the receiver asked for the hemisphere and recorded it in its memory, but the receiver does not display the hemisphere on the screen. The "N" in the second coordinate indicates that the coordinate lies in the northern hemisphere. A letter "S" appearing in the same position shows that the location is below the equator. In the third coordinate, the letter "U" is borrowed from the Military Grid Reference System (MGRS is described in Chapter 13) and specifies position relative to the equator.

The MGRS divides each UTM zone horizontally into 8° sections and assigns letters as shown in the figure opposite. The letter "U" means your position lies somewhere between north latitude 48° and 56°. The word north starts with "N," which will remind you that in the MGRS system the letter "N" and every letter after it specifies a location above the equator. Do not mistake the letter "S" for the southern hemisphere if your receiver uses the MGRS letters.

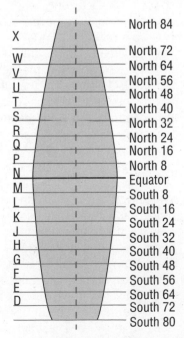

X	North 84
W	North 72
V	North 64
U	North 56
T	North 48
S	North 40
R	North 32
Q	North 24
P	North 16
N	North 8
M	Equator
L	South 8
K	South 16
J	South 24
H	South 32
G	South 40
F	South 48
E	South 56
D	South 64
	South 72
	South 80

MGRS letters applied to a zone.

UTM Rulers

As mentioned earlier, UTM coordinates on a large-scale map can easily be read by the unaided eye. However, a UTM grid ruler makes an already easy-to-use grid even simpler. UTM grid rulers are available through a variety of sources accessible over the Internet. A sample UTM grid ruler is shown in the adjacent figure.

To measure a coordinate, place the corner of the ruler at the location whose coordinate you want to measure. The lines of the UTM grid intersect both the vertical and the horizontal scales on the ruler. The figure shows how to measure the UTM coordinate of a building on the USGS Long Branch, NJ, topographical map. Note that the horizontal ruler scale is intersected by an easting grid line. The easting coordinate is found by adding the number where the grid line crosses the ruler scale to the base number of the easting grid line. In this case the ruler scale is intersected at 480, so the easting coordinate becomes:

$$575000m.E.+480 = 575480m.E.$$

A northing grid line intersects the vertical ruler scale. The northing coordinate is found the same way as the easting coordinate above. In this example, the vertical ruler scale is crossed at 650 by the 4458000m.N. grid line. The northing coordinate becomes:

$$4458000m.N.+650 = 4458650m.N.$$

The final complete coordinate is:

18 T 575480m.E. 4458650m.N.

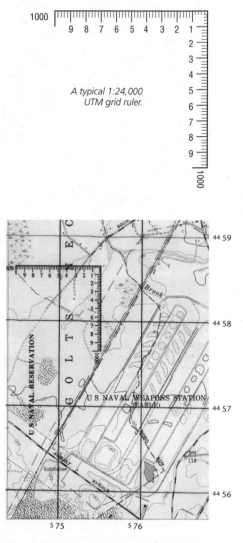

A typical 1:24,000 UTM grid ruler.

Long Branch, NJ USGS Topo.

The coordinate of any location is measured in the same way.

GPS Navigation in a Whiteout

Another practical use of GPS technology is as a backup means of navigation in poor weather. Let's say you plan to take some friends to the top of Mount Columbia, located in the Canadian Rockies. To get to the peak, you need to cross the Columbia Icefields, which lie at an altitude of 3038 to 3353 m (10,000 to 11,000 ft.) and like any high mountain area they are subject to whiteout conditions that make navigation extremely difficult. Fortunately, clouds do not affect satellite signals, so this trip is a perfect application for a GPS receiver.

Most people who climb Mount Columbia set up a base camp on the northeast side of the icefield. If the weather is bad, they stay in camp and wait it out. If the weather is clear, they carry only their survival gear and hurry across the icefield to the peak. However, even if the weather is clear during the trip to the mountain, it is not unusual for clouds to come up out of the valleys very quickly, resulting in whiteout conditions for the return trip. Preparations for finding your way back in conditions of poor visibility are made before starting across the icefield by sticking bamboo wands in the snow perpendicular to the proposed return route.

The normal procedure for navigation in a whiteout, without a receiver, is to use a compass and dead reckoning. It is no easy task to travel seven to eight kilometers (close to five miles) across an almost flat icefield in zero visibility. If your navigation is accurate, you cross the icefield, hit your wands and follow them back to camp. Navigation can be so difficult that parties have been forced to spend a cold night dug into the snow while they waited for the morning sun to burn off the cloud.

The first day of the trip, you will ski up the Athabasca Glacier to the Columbia Icefield via a snow ramp up a headwall between large crevasses and set up base camp. The camp is located where it is easy to find the top of the ramp even in poor weather. The second day, you will cross the flat, featureless icefield to the base of the summit, where you leave your skis and climb to the top. On the way to the peak, you have to cross a "trench" in the icefield formed where two heavily crevassed glaciers drop away on each side. In order to safely cross, it is critical to find the highest point of the trench where the crevasses are the smallest. You have made the trip several times before and are intimately familiar with the route, so you do not enter waypoints in advance. However, you will store critical points during the trip to the peak, just in case you need to use the receiver to retrace your steps in bad weather.

The day of the trip arrives and after a long drive you reach the staging area. You carry a map with you even though you are familiar with the route, and of course you have your compass just in case the receiver fails to perform. Before you leave your vehicle, you initialize the receiver to the following settings:

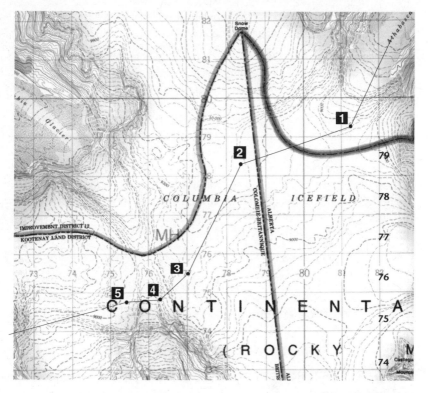

Map of the Columbia Icefield.

- **Map Datum: North American Datum 1927 (NAD 27)**
  If your receiver splits NAD 27 into separate settings for Alaska, Canada, Central America, etc., select **NAD 27–Canada;** otherwise select **NAD 27.**

- **Units: Metric**
  The UTM grid is based on the meter, so it is much easier to use the map if the distance is set to metric units.

- **Coordinate Grid: UTM**

- **North Setting: Magnetic North**
  A compass may be used to get between several points. If the receiver is set to report direction as magnetic bearings, the bearings may be directly dialed into a compass without compensating for the declination.

- **CDI Limit: Small**
  If the Course Deviation Indicator's tolerance is selectable, set it somewhere between 250 and 500 m. A CDI limit of 250

to 500 m means that at most you will stray 250 or 500 m off course before the receiver warns you.

- **WAAS Setting: Off**
  WAAS does not provide coverage in this area, so turn WAAS off so as to not introduce error.

After you initialize your receiver, you return it to your backpack and clip on your skis. You do not bother to mark the car's location, because there is no difficulty finding your way down the Athabasca Glacier to the staging area, even in horrible weather. You and your friends start off and in no time reach the top of the ramp at the headwall, which is the first point you need to record. You need a breather here anyway. You get the receiver out of your backpack and after it locks onto the satellites, you record the first point of the trip:

#1   11U   481019m.E.   5779434m.N.   Ramp

From the top of the ramp, you ski onto the Columbia Icefield to a flat area just south of the Snowdome. You set up camp and place a line of wands for tomorrow's return trip, as a safety precaution. You also record the position of your camp:

#2   11U   478327m.E.   5778541m.N.   Camp

The rising sun finds you and your friends equipped and starting across the icefield. The weather is beautiful. As a precaution, you have an extra set of batteries in your backpack. If the weather turns bad and you need to use your receiver to get back, you will have to use the energy-eating backlight to be able to see the receiver's screen. You also carry an extra bundle of wands. If the cloud descends, the receiver will get you to the best part of the trench and maybe exactly where you need to cross, but you will place wands at both ends of the trench and along the best passage to make the crossing easier in whiteout conditions.

The second day is going as planned and you make rapid progress to the top of the icefield, where you make a right turn to head to the trench. You take a short break and record the turning point as another waypoint:

#3   11U   477016m.E.   5775836m.N.   R Turn

Soon you reach the trench, where you spend some time finding the narrowest part. Once you find the best spot, you strategically place wands and mark it as another waypoint:

#4   11U   476328m.E.   5775207m.N.   Trench E

As you cross the trench, you continue to place wands to indicate the best passage. The receiver can lead you close, but the wands provide the increased accuracy necessary for a safe crossing. The important function the receiver performs is to get you close enough to the trench that you can find the wands – something you would not dare try with compass alone. Once on the other side of the trench, you record another waypoint:

#5   11U   475542m.E.   5775023m.N.   Trench W

Now both the east and west sides of the trench are marked. If a whiteout occurs, the receiver will be able to help you enter and cross the narrowest part of the trench. You have to travel 5 km (3.1 mi.) and an elevation gain of 610 m (2001 ft.) before you reach the mountain base, so you put the receiver away and start off again. When you reach the base of Mount Columbia, you leave your skis and mark their position:

#6   11U   470518m.E.   5774345m.N.   Base C

You climb the last 370 m (1214 ft.) to the summit pyramid, taking care not to fall through the double cornice on the summit ridge. At the top, you enjoy the marvelous view of the surrounding mountains, with Mount Robson a glistening white fang to the north. You take the receiver out of your backpack and mark the summit – not that you will need it for navigation, but as a memento of being at one of the most gorgeous places on earth.

You decide to make a route of the return trip, just to see the distances and bearings between the waypoints. The map shows all the waypoints except the last two. The information

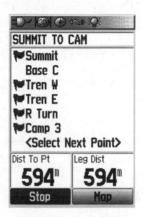

Your return route in list form.

given from the route function is shown below. The distances are in kilometers and the bearings have a magnetic reference.

Name	Point	Desired Bearing	Distance km
Summit	#7		
		69°	0.6
Base C	#6		
		63°	5.1
Trench W	#5		
		58°	0.8
Trench E	#4		
		29°	0.9
R Turn	#3		
		7°	3.0
Camp	#2		
		53°	2.8
Ramp	#1		

The receiver adds up the distances of the journey's legs from Summit to Camp to get 10.4 km (6.5 mi.). You never really knew the exact distance from the base camp to the peak.

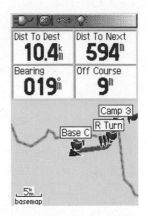

Looking at the long journey ahead, you quietly daydream on how nice it would be if you could just sprout wings and fly to the camp. Pleasantly engaged in this wistful fancy, you activate the Goto function and enter Camp as the destination. The receiver reports that the flight distance from Summit, your present location, to Camp is a distance of 9.4 km at a bearing of 41°. You cancel the Goto function. It seems that growing wings, and whatever that might entail, would save only 1 km. You decide to ski back to camp.

The route from Summit to Camp as displayed on the map.

Your fanciful flight path would save only 1 km.

When you look up from the receiver, you notice the cloud from the valley below creeping up the glacier toward the icefield. It has already filled the trench, which means it is time to get moving – fast. You put the receiver into your backpack and start down to the base with your friends. By the time you reach your skis, you are in thick cloud and the wind is filling your upward track with blowing snow, but you are not worried. With the receiver guiding the way, it will be a record run back to camp and not a cold night out.

Once your skis are on, you pull out the receiver, get it locked, turn on the backlight for

trench

easier reading in the poor light and press the Goto function. You need to get from the base to the west end of the trench, so you select Trench W as the waypoint destination. You select the Compass steering screen that reports your current heading and the necessary heading to reach the trench. It will be just like using a compass as you have done several times in the past, but better because you will always know the correct bearing to get to your destination even if you move off course. The steering screen shows the trench at a bearing of 63°, so you set course (heading) to match the bearing.

You look at the estimated time en route (ETE) on the screen (Time to Destination). At your present speed and if you stay on course, it will take about 30 minutes to get to the trench, but you know you will have to slow down, so it will take even longer. Note: if you had activated the Route function, as opposed to the Goto function, the Time to Destination would show the estimated time until your arrival at the Camp waypoint, Time to Next would show the time for each individual leg in the route.

As you cautiously ski toward the trench, you find you can stay on course reasonably well, so you turn off the backlight and only turn it on for an occasional quick look to make sure you are still on course. It seems as though your plan to get back to camp using the receiver will work, but only if you have enough batteries. Even though you have some spares, you know you had better conserve.

Each time you look at the steering display, if your present heading does not match the correct bearing, you check the Off Course statistic, whether on the Compass screen or the Highway steering screen, to see how far off course you are. During one such check, you see you are

Mt. Columbia

79 m (259 ft.) to the right of the direct line to Trench W, and the Highway is pointing to the left telling you to steer left to return to course. You are ecstatic about the news of being off course by so little because, before owning a receiver, you had never stayed on course so closely in a whiteout. Even better than staying on course, you know exactly how far you have strayed and what to do to return to course. You adjust the direction you are heading from 88° to 62°, which is 26° to the left, and continue on.

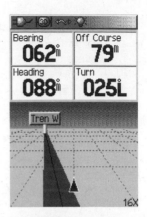

The next time you check the receiver, you notice you have strayed again, but this time you are to the left of the direct course by 22 m (72 ft.).

As you have been off course for a while, the bearing to Trench W has changed from 63° to 62° and now to 64°, so you steer slightly to the right to follow the new bearing. You change to the Compass screen and see that there remains just over 2.4 km (1.5 mi.) to the trench – something you would never have known without a GPS receiver. You press on, making occasional course adjustments until finally, two hours after leaving the base of Mount Columbia, you arrive at the west end of the trench and find your wands.

You turn off the receiver while you rope up and prepare to cross. When you are ready to go, you turn the receiver on and activate the Goto function. You select the Trench E waypoint. The steering screen shows the bearing to be 53°, so you leave the receiver on and watch the screen as you adjust your course to match the bearing. As you enter the trench, your pace really slows and you notice the receiver no longer gives you an ETE, but you know it can still accurately calculate the bearing from your present position to the other side of the trench. This juncture in the journey is critical, so you leave the backlight on and watch the screen continuously while feeling for wands. Occasionally you check the Map screen to verify your overall position in the trench and your course.

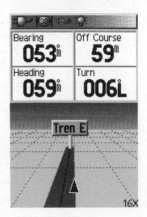

As you progress, the Map screen shows that you drift between left and right of the direct course. Each time, you correct your direction of travel to try to stay on the straight-line course to the Trench E waypoint. Your confidence in the receiver increases when you hit your first line of wands.

You change again to the Highway steering screen and see that you are off course to the right by 59 m (193.5 ft.). You are farther off course than you like or need to be and part of the problem is your own lack of vigilance. The receiver reports that your current direction is 59°, but you need to travel a bearing of 53° to get to Trench E. Switching back to the Compass screen you see that after traveling 9 meters to your left (toward the straight-line course) the bearing from your present position and the "To Course" heading have become the same (53°), and that continuing in this direction will take you to the destination.

You continue walking, paying more attention to the receiver screens to stay on course. A slow hour later, you find all the wands you placed and the receiver's map indicates that you have successfully made it through the trench to the Trench E waypoint. This is a first. You have never gotten so far so fast in such poor visibility. You would not have been able to do it without the combination of the receiver and wands. Now it is nearly a straight shot back to camp. You take off your ropes and activate the Goto function to lead you to R Turn.

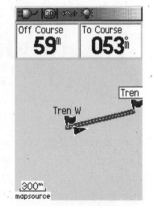

A bearing of 53° returns you to the direct line course.

The "To Course" heading and the Bearing converge.

Your vigilance on the leg to R Turn results in a relatively quick trip of 40 minutes. After resting, you change the batteries and activate the Goto function to lead you to Camp. Watching the receiver, you adjust your heading to 7°. The receiver calculates an ETE of approximately one hour at your present speed. With a bit of weaving back and forth to stay on track, you hit your wands an hour and fifteen minutes later. Yeah! It is going to be a warm night in the tents. You follow the wands back to camp and as you unzip the flap to your tent, you discover someone else is already in it! They got off course, hit your wands and are taking refuge from the weather in your tent. You make a note in your journal: "Bring extra receivers to sell at the staging area."

7 Latitude/Longitude, and a Kayak Trip

The Latitude/Longitude Grid

The latitude/longitude grid is familiar to most people. It is printed on almost all maps even if it is not the primary grid. If you travel in a part of the world where your receiver does not have the local grid, do not worry, because the map probably has latitude and longitude. The latitude/longitude grid is based on a sphere. The figures shows how the globe is divided by the lines of latitude and longitude.

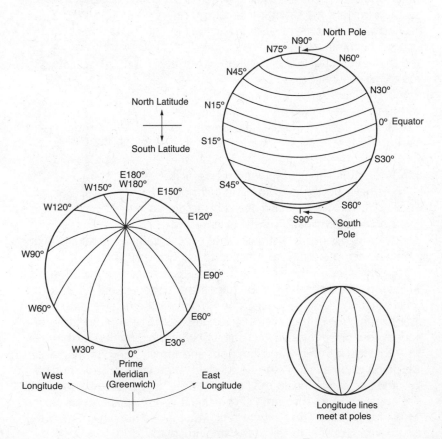

Lines of latitude go around the world parallel to the equator. Lines of longitude go from pole to pole. On a map, latitude lines are horizontal and longitude lines are vertical. Coordinates in latitude/longitude are expressed as degrees as in the degrees of the circle of the globe. Here are some interesting facts about latitude and longitude.

Latitude

- Lines run parallel to the equator.
- Hemispheres designated North (N) and South (S):

  Equator is 0°.

  North pole is N 90°.

  South pole is S 90°.

- May be expressed in three formats:

  Hemisphere, degrees, minutes, seconds: S 38° 27'54"

  Hemisphere, degrees, minutes: N 23° 27.3'

  Hemisphere, degrees: N 58.385°

- 1° of latitude is 111.12 km (69.05 mi.).
- 1' is one nautical mile (1.85 km, 1.15 mi.).

Longitude

- Lines run from pole to pole.
- Greenwich, England, is the prime meridian.
- Hemispheres designated West (W) and East (E) of the prime meridian:

  Prime meridian is 0°.

  International date line is W 180° (same as E 180°).

- May be expressed in three formats:

  Hemisphere, degrees, minutes, seconds: E 140° 54'09"

  Hemisphere, degrees, minutes: W 67° 28.75'

  Hemisphere, degrees: E 86.824°

- Longitude lines converge at the poles.
- 1° of latitude is 111.12 km (69.05 mi.) only at the equator.
- 1' is not one nautical mile except at the equator.

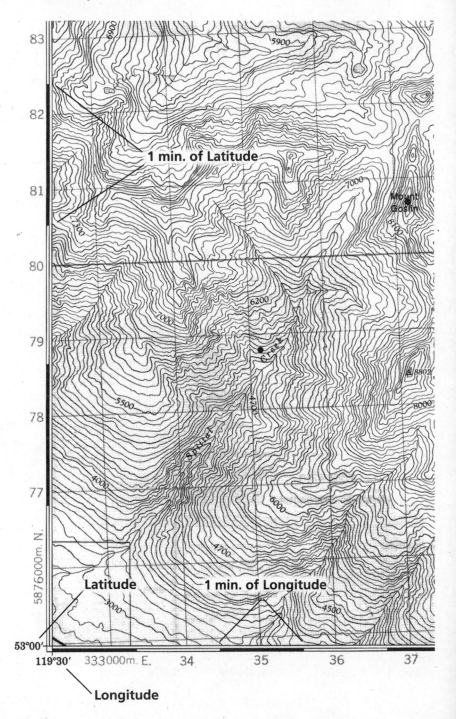

Latitude and Longitude Coordinates

* Coordinate usually written hemisphere latitude, hemisphere longitude.

   N 47° 19.56', E 102° 42.84'

   Called geographic coordinate because it is based on a sphere.

Finding the Latitude/Longitude Grid on Maps

Some maps make it easy to use the latitude/longitude grid. The map shown, a product of Natural Resources Canada, is the southwest corner of the Mount Robson, British Columbia, Canada, topographical map. It has a 1-minute latitude/longitude grid.

The base coordinates for the map are shown in the bottom left corner. For this map, the base latitude is N 53° 00' while the base longitude is W 119° 30'. From the corner, the latitude/longitude grid is marked off in one-minute intervals as indicated by the alternating black and white lines printed on the edges of the map. Also visible on the map's edges is the corresponding UTM grid information, which has nothing to do with finding latitude/longitude coordinates.

A latitude/longitude coordinate is found by adding the number of black and white lines between a location and the base coordinate. For example, on the left side of the map, the end of the white line, just below the 77 of the UTM grid, represents the latitude N 53° 01' (53° base + 1 minute). The end of the black mark above that is N 53° 02' (53° base + 2 minutes), and so forth. The longitude is measured the same way, but the minutes decrease as you travel east because you are traveling toward the prime meridian. The end of the first black line at the bottom of the page is W 119° 29' (199° 30' base – 1 minute); the end of the next white line is W 119° 28' (199° 30' base – 2 minutes), etc.

For either latitude or longitude, one-half of an alternating line is one-half of a minute, one quarter is one-quarter of a minute, and so forth. To determine the coordinate of a location, place a ruler perpendicular to the side of the map for latitude or to the bottom for longitude and measure to the closest tenth of a minute. Do not be confused by the UTM grid, which is also printed on the map; just ignore it. Two examples of coordinates taken from the map on the facing page are:

* Mount Goslin:
   N 53° 3.2' W 119° 25.7'

* The Letter "C" in Spittal Creek:
   N 53° 2.0' W 119° 27.4'

It is possible to measure coordinates with your eyes alone on a map with such a clear and easy-to-use system as the one shown in the picture. Other maps require some preparation to use the latitude/longitude grid.

Preparing USGS 7.5' Maps

The 7.5' topographical maps printed by the U.S. Geographical Survey require some preparation when using the latitude/longitude grid. Tick marks on the side of the map and crosses inside the map, as shown in the figure to the right, divide the map into 2.5' rectangles. To draw the grid, use a ruler to draw from a latitude/longitude tick mark on one side of the map through the crosses to the corresponding tick mark on other side. A subdivided map looks similar to the figure.

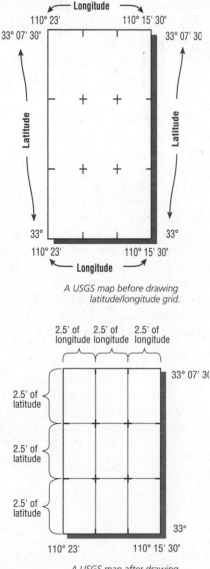

A USGS map before drawing latitude/longitude grid.

A USGS map after drawing the latitude and longitude grid.

A 2.5' rectangle of the southwest corner of the USGS Burrows Lake, Wisconsin, topographical map on the next page demonstrates that an area 2.5' × 2.5' is much too large to accurately read a coordinate without taking further steps. Using your eyes alone, try to find the coordinate of the island in the southern part of Burrows Lake, labeled #1.

The longitude coordinate for Burrows Lake is just over halfway between W 89° 50' 00" and W 89° 52' 30", which makes it about W 89° 51' 15". The latitude coordinate is about one-fifth of the way between N 45° 37' 30" and N 45° 40' 00", which gives it a latitude about N 45° 38' 00". The island's actual coordinate is N 45° 37' 55" W 89° 51' 23", which means the visual measurement differed from the true coordinate by 8" in longitude and 5" in latitude. An error of 5" in latitude equates to 0.083', which is the same as 0.083 nautical miles or 504.32 ft. (153.7 m).

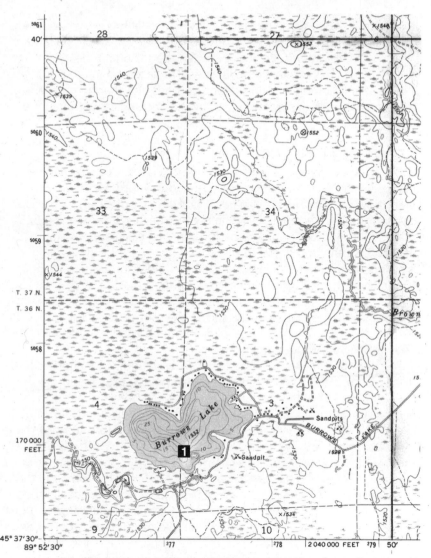

A 2.5' area on USGS Burrows Lake, Wisconsin, map (1:24,000 scale).

The 8" error in longitude does not directly correlate to nautical miles, since the latitude of Burrows Lake is not at the equator, but the distance scale at the bottom of the map reveals that 2.5' of longitude at this latitude is 2 statute miles. Once all the conversions are done, 8" in longitude on this map at the latitude of Burrows Lake is an error of 563.2 ft. (171.6 m). The error was not really that bad for an eyeball-only estimate, but it is much easier to use latitude/longitude on a USGS 7.5' map if the 2.5' grid is further subdivided.

There are two easy methods to subdivide any latitude/longitude grid that is too large for convenient use. The first is to use an ordinary ruler and a pencil to draw more lines between the main latitude/longitude lines, which in this case are the 2.5' lines. The second is to use a special ruler that is calibrated to minutes and seconds and that permits direct reading of coordinates from the 2.5' by 2.5' rectangles. Both methods will be illustrated.

Drawing a Finer Grid

The key to subdividing a latitude/longitude grid is to divide it by a number that results in a fraction that is easy to add in your head. It is easy to divide any grid into four equal parts, but in the case of a USGS 7.5' map, dividing the 2.5' rectangle by 4 means that each subdivision is 0.625' or 37.5" wide. It is not easy to perform mental calculations with units of 0.625, so the grid needs to be divided more sensibly. The 2.5' grid is best divided by either 5, which results in subdivisions of 0.5' each, or 10, which provides 0.25' between lines. It is the easiest to work with 0.5' intervals because there are fewer lines to draw, so 5 is probably the best divisor to use. The map opposite shows the southeast corner of USGS Burrows Lake with the 2.5' rectangle subdivided into five equal parts in both latitude and longitude.

The finer grid makes it much easier to find latitude/longitude coordinates accurate to about a tenth of a minute. For example, using the finer grid, the "S" in the northern part of Swamp Creek, label #1 on the map, has a latitude that is two-fifths of the way between N 45° 39.5' and N 45° 40', which is N 45° 39.7', and a longitude four-fifths of the way between W 89° 46.5' and W 89° 47', which is W 89° 46.9'. The exact coordinate for point #1 is N 45° 39.63' W 89° 46.85'. The resulting error in latitude is 0.07' (4.2") or about 425.3 ft. (129.6 m) and the error in the longitude measurement is 0.05' (3"), which for the latitude of this map equates to about 211.2 ft. (64.4 m). The measurement is only slightly more accurate that the example of Burrows Lake, where only the 2.5' grid was used, but the finer grid used in this example makes it so much easier to find a coordinate using the eye alone. The exact coordinates of the four locations marked on the map are given below. See how close you can come to getting the same coordinate without using a ruler.

Point	Latitude	Longitude
#1	N 45° 39.63'	W 89° 46.85'
#2	N 45° 38.68'	W 89° 47.16'
#3	N 45° 37.83'	W 89° 46.11'
#4	N 45° 39.36'	W 89° 45.66'

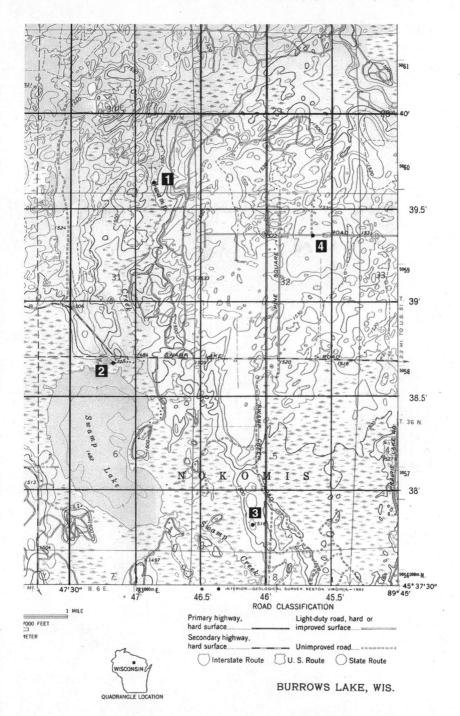

ROAD CLASSIFICATION

Primary highway,
hard surface.......................

Secondary highway,
hard surface.......................

Light-duty road, hard or
improved surface..............

Unimproved road............

◯ Interstate Route   ◻ U. S. Route   ◯ State Route

1 MILE

7000 FEET

1ETER

WISCONSIN

QUADRANGLE LOCATION

BURROWS LAKE, WIS.

105

Using a Minute/Second Calibrated Ruler

The easiest and most accurate way to measure latitude/longitude is to use a minute/second calibrated ruler. Such a ruler provides a method to measure coordinates to an accuracy of about 1" on a 1:24,000 scale map, which translates to an accuracy of approximately 30.9 m (101.3 ft.) or better. The time needed to prepare a map to use a minute/second calibrated ruler is minimal. On a USGS 7.5' map, using a minute/second calibrated ruler requires that you draw only the lines for the 2.5' rectangles as described earlier. The ruler does the additional subdividing to make reading coordinates simple and accurate. An example of such a ruler is the Topo Companion shown here. If it is difficult for you to find the Topo Companion, you can make your own minute/second calibrated ruler as shown in Chapter 8.

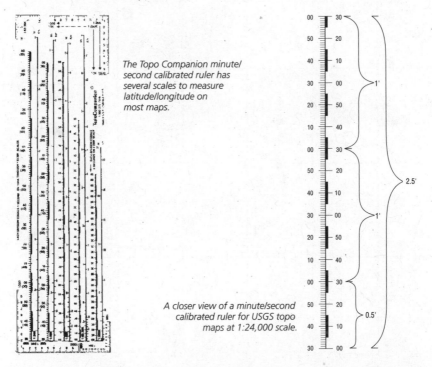

The Topo Companion minute/second calibrated ruler has several scales to measure latitude/longitude on most maps.

A closer view of a minute/second calibrated ruler for USGS topo maps at 1:24,000 scale.

The numbers along the bottom of the Topo Companion ruler correspond to map scales 1:24,000, 1:25,000, 1:50,000, 1:63,000 and 1:250,000. The USGS 7.5' topographical maps use the 1:24,000 scale on the minute/second calibrated ruler. The numbers on the 1/24k section of the ruler represent tens of seconds while the lines between the numbers are individual seconds. How the numbers correspond to the 2.5' rectangle is shown in the figure. The ruler measures a full minute whenever it starts at a given number and ends on a number of the same value. The figure below shows a full minute from 00 to 00 or 30 to 30. However, a full minute is also traversed going from 10 to 10 or 20 to 20, and so forth.

Note that the ruler starts at the bottom with 30 on the left and 00 on the right. The latitude or longitude of a 2.5' can start with either 00" or 30". The latitude of the southeast corner of the USGS Burrows Lake is N 45° 37' 30" while the longitude is W 89° 45' 00" (the 00" is not printed on the map, but it is shown here to demonstrate that it corresponds to the 00 on the minute/second calibrated ruler). If the coordinate ends on 30", use the numbers that start at 30 in your measurement. If the

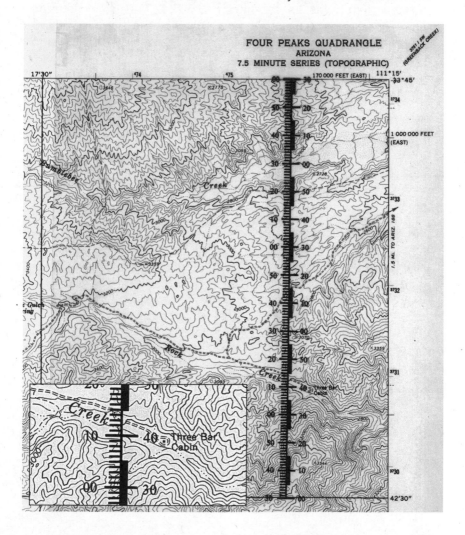

coordinate starts at 00", or if the seconds are omitted because they are zero, use the numbers that start with 00 when measuring.

The key to using a minute/second calibrated ruler is to place the ends of the scale on adjacent latitude or longitude lines, then count off the seconds to the location. The map shown on the preceding page is the northeast corner of the USGS Four Peaks, Arizona, map.

The lower horizontal and left vertical drawn lines combined with the top and right edges of the map form the 2.5' rectangle. To measure latitude, one end of the ruler touches the bottom of the 2.5' rectangle while the other end touches the top. From the bottom of the ruler, count up the number of seconds to the location of Three Bar Cabin. The latitude of the bottom of the 2.5' rectangle is N 33° 42' 30", so when using the ruler refer to the numbers on the left of the scale that start with 30. The numbers on the left side of the scale and how they correspond to a latitude coordinate on the map are given below:

Number on scale		Corresponding latitude
30	→	N 33° 42' 30"
40	→	N 33° 42' 40"
50	→	N 33° 42' 50"
00	→	N 33° 43' 00"
10	→	N 33° 43' 10"

Three Bar Cabin lies above 00, but below 10, so its latitude coordinate is between N 33° 43' 00" and N 33° 43' 10". The small tick lines between 00 and 10 must be counted to get to the exact coordinate. Each tick mark is one second. The cabin lies directly across the ninth tick mark, so the final latitude coordinate is N 33° 43' 09".

It is just as easy to measure the longitude coordinate. The figure opposite shows how the ends of the ruler are placed on the longitude lines of the 2.5' rectangle. Notice that the ruler is not completely horizontal, because only at the equator is 2.5' of latitude the same physical distance as 2.5' of longitude. However, even though the ruler is at an angle, if the ends of the scale are on the longitude lines, the measurement will be accurate. Remember, the ruler measures minutes and seconds, not distance, so the spacing between the longitude lines is unimportant. The only requirement for the ruler to work on any scale is that the number of minutes and seconds between adjacent latitude and longitude lines be the same. In this case, there are 2.5' between the latitude lines and the same amount between the longitude lines.

From the right-hand end of the ruler, count up the minutes and seconds to the point directly above Three Bar Cabin. The coordinate of the longi-

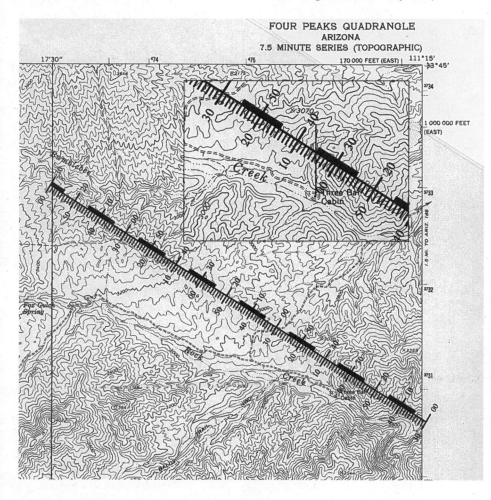

tude line on the right of the 2.5' rectangle is W 111° 15', so when measuring, use the numbers on the right, or in this position the top, of the scale,
because they start with 00. The relationship between the numbers on the
scale and the corresponding longitude coordinate is given below:

Number on scale		Corresponding longitude
00	→	W 111° 15' 00"
10	→	W 111° 15' 10"
20	→	W 111° 15' 20"
30	→	W 111° 15' 30"
40	→	W 111° 15' 40"

109

The location of Three Bar Cabin lies between 30 and 40 on the scale, which translates to a coordinate between W 111° 15' 30" and W 111° 15' 40". Count the small tick marks between 30 and 40 to get the exact coordinate of W 111° 15' 35". The final coordinates for Three Bar Cabin in latitude/longitude are:

N 33° 43' 09", W 111° 15' 35"

A Kayak Trip

In May, your brother is coming to visit you in Dupont, Alaska, and really wants to go to your cabin. When you describe how beautiful and peaceful it is there, he decides he wants to go as soon as he arrives and stay for the entire three days he will be in town. The only transportation you have between your house and the cabin is a kayak. It is only a 10.8 mi. (17.4 km) paddle, but your brother's plane arrives in Juneau at 10:30 pm and it will be dark before you can even start the trip. It is easy to navigate to the cabin by sight during the day, and in the dark it is still possible, but much more difficult.

You decide your GPS receiver can guide you directly to the cabin, even in the dark, if you set it up so the navigation screens are continuously visible. It is impossible to hold the receiver and paddle at the same time, so you mount the receiver to the kayak. Another problem you need to overcome is powering the receiver. You will make the trip in the dark and will need the receiver's backlight continuously illuminated so you can see the screen. You plan to take extra batteries and will put in a fresh set if necessary. You also need to attach lights to the kayak to paddle safely through the busy traffic of Stephens Passage. You decide to use the Route function to automatically guide you from one waypoint to the next without touching any buttons, thereby leaving your hands free to paddle.

Using your minute/second calibrated ruler, you measure the latitude/longitude coordinates for the trip's waypoints from the USGS Juneau, Alaska (A-1) map. The map is from the USGS 15' series and has a scale of 1:63,360. The latitude/longitude grid on the map forms 5' rectangles instead of the 2.5' rectangles of the 7.5' maps described previously. The minute/second calibrated ruler is used in exactly the same way as demonstrated previously except you use the 1:63,360 scale ruler to make the measurements.

The coordinates and names of the five waypoints marked on the map opposite are shown on the page following the map.

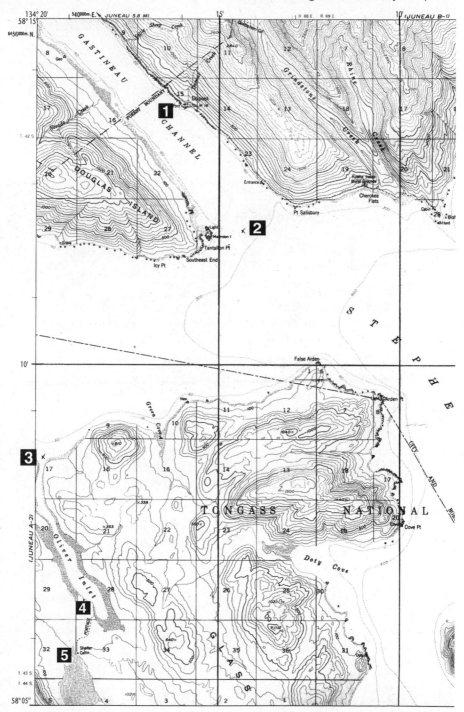

USGS Juneau, Alaska (A-1) (1:63,360 scale).

Point	Latitude	Longitude	Name
#1	N 58° 13' 45"	W 134° 15' 55"	Dupont
#2	N 58° 11' 53"	W 134° 14' 20"	Tantal
#3	N 58° 08' 40"	W 134° 19' 45"	Oliver
#4	N 58° 06' 35"	W 134° 18' 35"	P Tage
#5	N 58° 05' 50"	W 134° 18' 47"	A Cabin

Before entering the coordinates into your receiver, set it to the proper settings:

- **Map Datum: North American Datum 1927 (NAD 27)**
  If your receiver splits NAD 27 into separate settings for Alaska, Canada, Central America, etc., select **NAD 27 – Alaska;** otherwise select **NAD 27.**

- **Units: Statute**
  Nautical units are best suited for latitude/longitude coordinates, but you have never used them before and cannot relate to a nautical mile, so use statute.

- **Coordinate Grid: Latitude/Longitude**
  Select the degrees, minutes and seconds format to correspond to the coordinates measured from the map.

- **North Setting: True North**
  A compass is not going to be used to navigate, so it does not really matter if the receiver's bearings are oriented to the magnetic or north pole. You arbitrarily decide to make the bearings relate to the map and select the True North mode.

- **CDI Limit: Small**
  If the Course Deviation Indicator's tolerance is selectable, set it somewhere between 0.25 and 0.5 mi. On the water, there are no obstacles in the route you have planned, so a small CDI will allow the Highway navigation screen to keep you on course.

- **WAAS Setting: On**
  As you are in an area where there is WAAS coverage, turn WAAS on for increased accuracy.

Once the waypoints are in the receiver's memory, you set up the route that will lead you from Dupont through all the intervening waypoints until you arrive at the cabin. The distance is expressed in statute miles and the bearings relate to true north as set above.

Name	Point	Desired Bearing	Distance mi.
Dupont	#1		
		156°	2.4
Tantal	#2		
		222°	5.0
Oliver	#3		
		164°	2.5
P Tage	#4		
		188	0.9
A Cabin	#5		

The plan for navigating the 10.7 mi. (17.4 km) is simple: use the receiver's Route function to guide you the entire trip. You will paddle the kayak from Dupont to P Tage, where you will portage the kayak the remaining 0.9 mi. (1.45 km) to the cabin. While you are in the kayak, you will have access to the steering screens and other navigational statistics like speed, bearing, estimated time en route, etc. Your receiver uses averaging and smoothing algorithms, so even at your slow speed the navigational statistics will be meaningful. You will also take a compass and a map, and if something were to happen to the receiver, you will use the compass to paddle due south to land, then either slowly follow the coastline west to Oliver Inlet or wait for sunrise to easily complete the trip.

You switch to the Map screen to view the route. The name of each waypoint appears on the screen. The straight-line route between each waypoint is shown. You move the map cursor arrow over the water. The name Stevens (Stephens) Passage identifies the body of water and shows the position of the cursor as N 58° 11' 12.8", W 134° 12' 31.7".

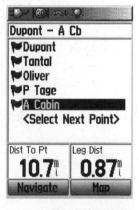

Creating your Route.

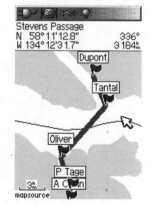

Viewing your Route.

Out of curiosity, you upload the waypoints to Google Earth. The resulting satellite image shows Dupont, the turning points Tantal and Oliver in Stephens Passage, Oliver Inlet with the waypoint P Tage at the end, and the cabin. Just as indicated on the topographical map, the mountains rise steeply from the water.

The day finally arrives. Your brother's plane lands at 10:40 pm and after several delays you are ready to shove off in the kayak at midnight. Before paddling, you turn on the lights and the receiver. Once the receiver is locked onto the satellites, you activate the route that leads to the cabin. The receiver instantly detects that you are already at the route's first waypoint, so it immediately starts pointing to the next waypoint, which is Tantal. The Compass screen shows the distance and bearing to Tantal as 2.31 mi (3.7 km) at a bearing of 156°. Your current heading is 154°, so you are headed slightly off course. The total distance from your current position to the A Cabin waypoint is 10.7 mi (17.2 km) along the selected route.

Directing you toward Tantal.

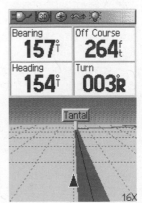

After paddling about halfway to Tantal, you switch to the Highway navigation screen and notice that you are slightly off course to the left of where you should be. You had been traveling at a bearing of 154° when it should have been 156°, and as a result you are off course to the left about 264 ft. (80.5 m) (CrossTrack Error), which is not bad. You also notice the bearing to Tantal has changed from 156°, as it was originally, to 157°. You do not really want to steer deeper into Stephens Passage, so you adjust your track to 157° and continue. As you hold your course, the distance you are off course steadily decreases as you approach the straight-line shown on the Highway screen. The receiver advises you of your arrival at Tantal, then automatically switches and starts steering you across Stephens Passage to Oliver Inlet. You adjust your course to the new bearing of 222°.

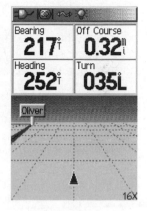

As you paddle along, you and your brother get talking and you do not pay attention to the receiver. When you do look up, you see you have gone almost 0.6 mi. (0.97 km) at a heading of 252°. The arrow on the Compass navigation screen points to the left and the bearing to Oliver is now 217°. You switch to the Highway screen. The direction of the Highway is off to the left, indicating you need to steer left to get back on course. It also shows you are off course by 0.32 mi. (1689 ft., or 515 m) to the right of the intended course.

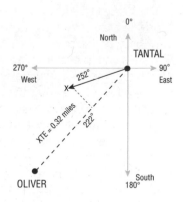

You decide to try an experiment to get back to the direct line between Tantal and Oliver as fast as possible. The adjacent figure illustrates the bearing between Tantal and Oliver is 222°, but because you paddled on a bearing of 252°, you are 0.32 mi. away from the direct line source as indicated by Cross-Track Error (XTE). The fastest way to return to the direct line is to follow the route indicated by the dotted line between the "X" that marks your current position and the line between Tantal and Oliver. The line between the X and the direct route is perpendicular to the direct route. The dotted line bearing is:

$$222° – 90° = 132°$$

You turn the kayak to the left until your bearing is 132° and paddle until the CrossTrack Error (Off Course number) is zero. Soon you are back on the direct route and you turn right to a bearing of 222° to head straight toward Oliver. Everything is going smoothly, without much traffic in the passage. However, when you get about halfway into Stephens Passage, you hear the loud sound of a fast motorboat. Fortunately they see you and give you plenty of room, but when the sound of the motor dies down, you hear something that sounds like a dog barking. No dog on land could sound that close, so you decide to investigate.

Before you start paddling toward the dog's bark, you activate the Man-overboard (MOB) function, which immediately marks your present position. You plan to investigate the sound, then return to where you are right now to continue the trip to Oliver Inlet. The MOB function not only marks your current position but also automatically activates the Goto function to steer you to the position just marked. While you are looking for the dog, you will ignore the receiver's screen. When you are ready to continue your trip, the receiver will lead you back to the MOB coordinate, where you will disengage the Goto function and resume the route to the cabin.

Bearing from Here to Oliver.

Out of curiosity you switch to the Map screen and see that the MOB waypoint is just less than halfway between Tantal and Oliver.

Because you are following the dog's bark and not the receiver, you leave it on the Map screen just to see where you are going in relation to the waypoints on the screen. As you paddle toward the bark, you notice that you are going all over the place and not in a straight line. Either the dog is afraid of you and is swimming away or you are having a hard time following the sound as it travels over the water.

Sure enough, the dog is not on land, and as you pull it aboard, it seems really happy to see you. When you notice it does not have any tags, you are pretty sure the owners will not even know where to begin to look for their lost pet. You take another look at the receiver Map screen. If you return to the MOB, it will take a lot of extra paddling because the dog was far away in a direction closer to Oliver than to Tantal. You decided the best approach to continue the journey is to paddle directly from your present position to Oliver, so you cancel the MOB Goto function.

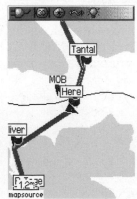

The edited route.

Your receiver can pick up a route where it left off, but you want to store your current position and include it into the route, so you store your current position as Here. You then insert the Here waypoint into the original route. The distance between each leg of the modified route is given below. You switch to the Map screen which shows the Here waypoint in the route and the MOB waypoint.

Name	Point	Desired Bearing	Distance mi.
Dupont	#1		
		156°	2.4
Tantal	#2		
		222°	2.4
Here			
		221°	2.5
Oliver	#3		
		164°	2.5
P Tage	#4		
		188°	0.9
A Cabin	#5		

From where you are right now, it is still 5.9 mi. (9.5 km) to the cabin. You activate the new route, switch to the steering screen and swing the kayak around to a bearing of 244°. The trip continues with a few course corrections until you arrive at the cabin.

Three glorious days later, when it is time to go home, it is easy to get the receiver ready for the trip back because, with a press of a button, the original route reverses. Then you discover that the Here waypoint is still part of the route. You select the route, select the Here waypoint, and remove it from the route. After deleting the Here waypoint, the receiver uses the route shown below to guide you home.

Name	Point	Desired Bearing	Distance mi.
A Cabin	#1		
		8°	0.9
P Tage	#2		
		344°	2.5
Oliver	#3		
		42°	5.0
Tantal	#4		
		336°	2.4
Dupont	#5		

Once more, in the dark with the receiver showing the way, you, your brother and your new dog paddle from Oliver Inlet back to the house, glad you were able to spend as much time there as you did.

8 More Latitude/Longitude, and a Sailboat Rally

Every year, your sailing club has a timed race where the object is not to be the fastest, but to be the closest to a set time between points. The time is fixed to teach you how to better control the boat, and you are docked points for every minute you are too fast or too slow. Points are also given for the proximity of your arrival at each location. Some locations are fairly remote and it would be difficult for judges to monitor a boat's arrival, so each crew is given a digital camera to prove how close they get to each marker.

It is your fifth year in the race. You finally understand how to handle the craft and can sail proficiently in most conditions, but you are hopelessly inept at judging your speed in the water. Last year, you zoomed as fast as you could to get close to each point, then waited until the time was almost up before moving in the last little bit. Your method left a lot to be desired because you placed 45th in a field of 100 contestants. This year you need to be a bit more controlled. You search the rules carefully and find there are no restrictions on GPS receivers. You outfit your boat with an external antenna, power cord and mounting hardware because you plan to use the receiver's navigational statistics to help you arrive right on time.

The stopping points of the race have been published on the official chart, so you reach for your ruler to measure coordinates for your route. The coordinate grid on the chart is latitude/longitude, so you grab your minute/second calibrated ruler to start measuring, but none of the scales seem to work. You notice the scale, at the bottom of the map, is 1:40,000.

"Great!" you complain. "The calibrated ruler you have does not have the scale you need. Then you notice the subdivided lines below the scale that have "LATITUDE" and "LONGITUDE" written above them.

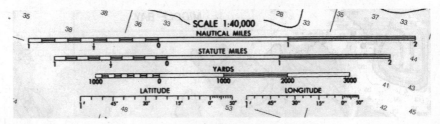

After comparing the lengths of the scales to the distance between the latitude and longitude grid lines, you realize that you can make your own minute/second calibrated ruler for the 1:40,000 scale. There are three ways to make the ruler:

1. Make one ruler, based on the latitude grid line, and use it with any 1:40,000 scale map, exactly like the Topo Companion ruler described in Chapter 7.

2. Make two rulers: one for latitude, the other for longitude. The longitude ruler would be the length between longitude lines for this map. The longitude ruler would be shorter than the latitude ruler (because the map area is well above the equator) and usable for this map only.

3. Make a ruler with longitude along the top and latitude along the left side as shown in the figure. This type of ruler makes it possible to place the ruler's corner on an object and easily read the coordinates. Once again, because the latitude ruler is based on the latitude spacing of the map, the ruler can only be used with this map.

You decide to use the third option. To make the ruler, first measure the distance between two latitude lines. For this chart at this scale, the distance is 9.2 cm (3.62 in.). Use centimeters because it is much easier to divide a length that is based on a factor of 10. Inches are subdivided into 1/4, 1/16 and 1/32 increments, which results in difficult math when dividing. Draw a vertical line 9.2 cm long on a piece of paper. Because the latitude lines are 2' apart, the line represents 2' of latitude. Label the bottom of the line 2' as shown in the figure. Measure 4.6 cm down from the top, or halfway down, draw another line and label it 1'. The top of this line represents 0', but do not label it because it will get in the way of the longitude line when it is drawn later. There is enough room between each mark to make 10 subdivisions, which translate into 0.1' subdivisions. The ruler shown in the figure has 0.1' subdivisions but only every other mark is labeled, so that the ruler does not look cluttered. A ruler that has only the latitude scale is equivalent to a Topo Companion ruler and can be used as shown

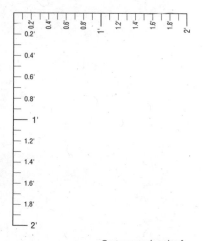

Custom-made ruler for a 1:40,000 scale chart.

earlier. But because you chose the option 3 ruler, you need to add the longitude scale.

Back on the chart, you measure 7 cm (2.76 in.) between adjacent longitude lines. On the paper you are using to make the ruler, draw a 7 cm horizontal line to the right from the top of the vertical latitude line. The horizontal and vertical lines meet a right angle. Just as with the latitude, the longitude grid lines are separated by 2'. Label the end of the line 2', draw a line at the halfway mark, 3.5 cm, and label it 1'. The space between the minute lines can again be subdivided into 10 equal parts to provide 0.1' marks. Once you have drawn the ruler, copy it onto a transparency so the map's features are visible when you use the rule to measure coordinates.

Measuring coordinates is easy with the ruler. The way the ruler is drawn requires the latitude scale to be always on the left and the longitude on top. To measure a coordinate, place the corner of the ruler on the location and note where the latitude and longitude grid lines of the map intersect the latitude and longitude scales of the ruler. Finding the coordinate to the mouth of Howells Creek is shown in the figure below.

The ruler's corner is placed at the mouth of the creek. The N 40° 44' latitude grid line intersects the ruler's latitude scale at 0.5'. The latitude coordinate is the sum of the latitude grid line and the value where the

US NOAA, Long Island Intra Coastal Waterway (1:40,000 scale).

grid line intersects the latitude scale. In this case, the latitude for the mouth of the creek is:

N 40° 44' + 0.5' = N 40° 44.5'

The same approach is used with the longitude. The W 72° 56' longitude grid line intersects the ruler's longitude scale between the 1.2' and 1.3' marks. Use your eye to estimate the distance between the two marks to arrive at the number 1.23'. Add the longitude grid line and the value where the grid line intersects the longitude scale. In this case, the longitude for the mouth of the creek is:

W 72° 56' + 1.23' = W 72° 57.23'

The coordinate to the mouth of Howells Creek is:

N 40° 44.5', W 72° 57.23'

Now that it is easy to measure coordinates from the map, you begin your work. This year's contest has four locations, numbered 1 through 4.

A quick look reveals that the path from #1 to #2 to #3 and #4 back to #1 are not straight lines. The receiver's Estimated Time En Route (ETE) calculation is based on the Velocity Made Good (VMG) measurement. Refer to Chapter 4 for an explanation of VMG and Speed Over Ground (SOG). If you are headed directly for a waypoint, the ETE is the actual time it will take to get there. If you are off course or the course is not a straight line, the VMG and ETE both vary widely. The only straight shot on the sailing course is between #3 and #4, so you break up the route between each point into a series of straight lines. You will use the receiver's Route function to lead you from one waypoint to the next, and if you stay on course the ETE will be the actual time between each point and you can use it to help meet the time requirements. You label the intermediate waypoints 1A, 1B, 2A, 2B and 4A.

The coordinates of all the waypoints you will use, along with their names, are listed below.

Point	Latitude	Longitude	Name
#1	N 40° 42.38'	W 73° 13.2'	P1
#1A	N 40° 39.22'	W 73° 12.6'	P1A
#1B	N 40° 38.71'	W 73° 11.2'	P1B
#2	N 40° 38.46'	W 73° 11.43'	P2
#2A	N 40° 39.2'	W 73° 10'	P2A
#2B	N 40° 39.73'	W 73° 10'	P2B
#3	N 40° 39.57'	W 73° 10.55'	P3
#4	N 40° 42.05'	W 73° 10.68'	P4
#4A	N 40° 41.6'	W 73° 12.2'	P4A

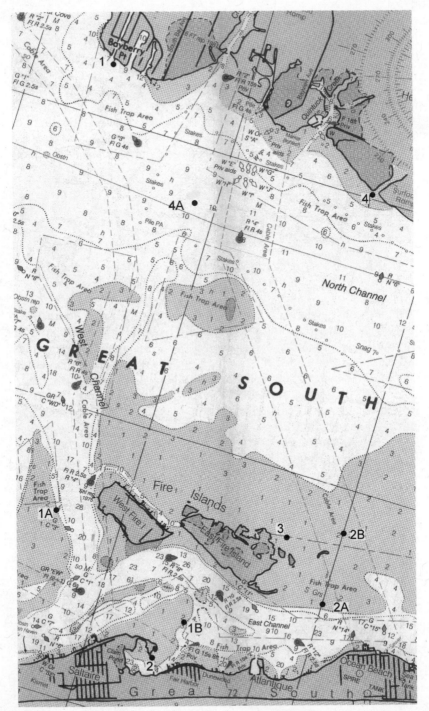

US NOAA, Fire Island Inlet (1:40,000 scale).

Before you enter the waypoints into the receiver, you need to specify the map datum. You search the official race chart in vain, but the datum is not included. You debate what to do. Never before have you had to know the datum for the race, because navigation with compass does not require it. If you ask the rally organizers, they will be suspicious and ask why you need it. You will have to make an educated guess. The chart was made in the U.S., so the datum is probably NAD 27, NAD 83 or WGS 84. If you pick the wrong datum, the position error will only be a few meters. This will not pose a problem, because you only have to get close to each point, not exactly on top of it. You decide to use NAD 27 because it is the datum used by most of the charts you own. Some of the newer charts use NAD 83, but you cannot tell from the copy if it is old or new. You set your receiver as shown below.

- **Map Datum: North American Datum 1927 (NAD 27)**
  If your receiver splits NAD 27 into separate settings for Alaska, Canada, Central America, etc., select **NAD 27 – CONUS**; otherwise select **NAD 27.**

- **Units: Nautical**
  You have quite a bit of experience with charts and have a feel for nautical miles, so select nautical units.

- **Coordinate Grid: Latitude/Longitude**

- **North Setting: Magnetic North**
  The big on-board compass is like an old friend, and even though you are using a GPS receiver, you will turn to your own compass when actually trying to maintain a course.

- **CDI Limit: A medium setting**
  You are proficient at holding an accurate course. The Highway navigation screen may prove useful, so you do not want a CDI limit that is so large that it is meaningless, or one that is so small that you can never hold the course. Your receiver has settings of 0.25, 1.25 and 5.0, so select 1.25 nautical miles.

- **WAAS Setting: On**
  As you are in an area where there is WAAS coverage, turn WAAS on for increased accuracy.

After setting the receiver, you enter the waypoints and then form a route.

Name	Point	Desired Bearing	Distance n. mi.
P1	#1		
		186°	3.2
P1A	#1A		
		129°	1.2
P1B	#1B		
		229°	0.3
P2	#2		
		69°	1.3
P2A	#2A		
		14°	0.5
P2B	#2B		
		263°	0.4
P3	#3		
		11°	2.5
P4	#4		
		262°	1.2
P4A	#4A		
		330°	1.1
P1	#1		

The total trip is 11.8 nautical miles (13.5 statute miles, 21.7 km). The time for each leg was also sent with the chart, so you quickly form a table listing times and distances, and you calculate the constant speed required to cover each leg in the specified time.

Forming the route "YachtRace" in the receiver.

Leg	Time min.	Distance n. mi.	Speed knots
#1 to #2	36	4.7	7.8
#2 to #3	21	2.2	6.3
#3 to #4	13	2.5	11.5
#4 to #1	24	2.3	5.8

You can easily sustain the 11.5 knots needed between #3 and #4, so none of the legs appear to be excessively fast. Next you calculate the time that can be spent on each leg as you defined them on the map. You decide that last year's strategy still has merit, but this year you will not rush too fast and then wait so long. You form the following table:

Leg	Distance n. mi.	Target Time min.	Speed Required knots
P1 to P1A	3.2	23	8.3
P1A to P1B	1.2	8	9.0
P1B to P2	0.3	5	3.6
P2 to P2A	1.3	11	7.1
P2A to P2B	0.5	4	7.5
P2B to P3	0.4	6	4.0
P3 to P4	2.5	13	11.5
P4 to P4A	1.2	11	6.5
P4A to P1	1.1	13	5.1

Fortunately you do not have to sail the craft solo. You will stay at the helm, keep an eye on the instruments, including the GPS receiver, and tell the crew what to do to keep on course and on schedule. You plan to use the Estimated Time En Route (Time to Next) timer on the receiver to calculate the time it should take to get to a given point. The ETE will help you sail the right speed to meet the time requirements of each leg. The ETE changes with speed and course. As you go off course, the ETE changes. The ETE on earlier receivers varied rapidly and significantly with only minor course deviations, but newer re-

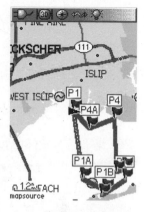

The route, as shown on the Map screen.

ceivers vary the ETE slowly to permit course corrections without major changes in the ETE. You decide to use a countdown timer to track the time elapsed on each leg. As long as the time on the countdown timer matches the ETE, you are on course and sailing at the right speed. You plan to use the countdown timer on your wristwatch.

On the day of the race, you have the chart, the tables you made, your receiver and your countdown timer. A boat leaves from the starting point every 10 minutes. As your turn to leave approaches, you turn the receiver on and activate the route. The steering page shows the bearing to P1A as 186°. When it is your turn, you shove off and keep an eye on the compass until you are sailing the bearing. According to plan, it should take 23 minutes to reach P1A. You set the countdown timer and activate it. You tell the crew to put on sail until you reach the desired speed of 8.3 knots. The ETE varies as compared to the countdown timer. You check your compass and adjust the helm to bring the ship closer to course. The countdown timer never quite matches the ETE, but they are close.

"Ship ahead!" one of the crew calls from the bow. You blast your horn because sailboats have right-of-way, but the other vessel does not move. You blast again, but there is no response, so you take evasive maneuvers. As you steer your boat hard to the left, the arrow on the Compass navigation screen points far to the right showing the direction you need to turn to get back on course. You switch to the Map screen. The triangle that represents your position and direction points away from the straight-line course between P1 and P1A. The bearing to P1A has changed from 186° to 193°. Once past the obstacle, you bring the boat around to a bearing of 193° to sail to P1A.

The countdown timer reports 9:30 to arrive, but the ETE is just over 15 minutes. You increase speed and sail at almost 12 knots until the two timers almost match. As you approach P1A, the receiver changes to steer you to P1B. You make a quick note of the time it took to get to P1A. With an eye on the compass, you bring the boat around to 129°. You made it to P1A in 22:47, which is almost perfect. According to plan, you reduce your speed to 9 knots to get to P1B. You also set the countdown timer to 8 minutes and start it.

"Cable ship ahead!" you hear from above. You are in a cable area, but you did not anticipate they would be working the weekend. They certainly will not move, so you had better do so. You veer hard to the right to get around the ship and then return to course. The countdown timer says 4:18, but the ETE is somewhere close to 6 minutes. More speed is needed, but there is another obstacle. A group of fishing boats are bringing in their nets, so you steer around them too. Once clear of all the obstacles, you have 2:08 to get to P1B, but the ETE at your present speed is over 5 minutes. You take a look at the Map screen. What a mess!

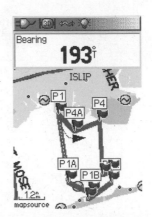

Turning to avoid the ship.

New bearing to sail toward P1A.

Steering around cable and fishing ships.

It takes 10:15 to finally get to P1B, which means you are 2:15 overtime. You can make up the time by going faster. You had allotted 5 minutes to get from P1B to P2, but with the overage, you now need to sail to P2 in 2:45. You set the countdown timer and bring the ship around to 229° to head for P2. It is a short distance and fortunately the way looks clear. You increase your speed to make up the lost time. You switch to last year's strategy of hurrying and waiting. When you arrive at P2, the countdown timer shows 30 seconds remaining. After you snap the photo of the race marker and get started again, you note it took 2:31 for the leg between P1A and P2, so even with the earlier delay, things are looking good.

The next leg should take 11 minutes. You reset your timer and assume a heading of 69° as you try to get up to the target speed of 7.1 knots as fast as possible. You arrive at P2A within the time allotted, but then you notice that there is more cable work going on in the area of P2B.

You decide to change the route to P3. Instead of sailing from P2A to P2B then to P3, you will sail from P2A to P3 and bypass P2B. You decide the best route is to sail between the small island near P2B and the main island. The time you allotted to get from P2A to P3 is 10 (6 + 4) minutes. As you are now going directly from P2A to P3, you have 10 minutes to do it. It is not a straight path to P3, so you will set the countdown timer and use the ETE as a guide, not as an exact measure. You will use the tactic of covering most the distance quickly and slowing down a lot once you are headed directly to P3 and the ETE can be compared against the countdown timer. You will ignore the bearing the receiver gives until you can go directly toward P3.

When you arrive at P2A, you start the timer, but you do not do anything to the receiver. Most receivers can detect when a waypoint is being skipped and will point to the next waypoint in the route. You cover most of the distance at about 7 knots. When you come close to P3, you have about 2 minutes to go. By now the receiver is pointing to P3 and you slow your speed so the ETE almost matches the countdown timer. A quick look at the moving map shows the detour you took.

Your plan worked out great because you reached P3 in 10:22. You take the photo of the marker, set the timer, change your heading to 11°, and try to pick up speed as fast as possible to reach your target of 11.5 knots. The total elapsed time from P1 to P3 is 56:50, while the allotted time is 57 minutes, so you are doing great!

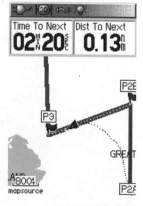

Path from P2A to P3.

The trip to P4 was not exactly straight because you had to avoid a few obstacles, but the receiver's ETE calculation and the countdown timer keep you on schedule and you arrive a little under at 12:25. You and the crew are working like a finely tuned machine, at least until you try to take the picture. In an effort to get as close as possible, you lean over the side of the boat. You snap the photo, but when you try to get back into the boat, your feet slip and you let go of the camera. You and the rest of the crew watch as it sinks to the dismal depths. You know the race is over – at least for you and your crew. Losing the camera means disqualification. Your time will not even be recorded.

You all decide to sail on as though nothing happened. Back at the helm you man the timer and receiver. It takes 11:48 to get from P4A and finally 13:18 to reach the finish line. Your total time is 94:21, which looks pretty good when compared against the 94 minutes set for the entire course. It is the closest you have ever come and if you had not dropped the camera the first prize might have been yours.

9 GPS Receivers and Personal Computers

If your current use of a GPS receiver does not involve a computer, it should. The most important feature of a computer program designed to work with a GPS receiver is the program's ability to transfer data to and from the receiver. The type of information transferred between the computer and the receiver depends on the type of program you buy and the capabilities of the receiver. There are three classes of programs available for use with a GPS receiver. Their capabilities and what they transfer to and from the receiver are described below.

Non-map
- Use to manage waypoints.
- Enter waypoints on the computer, transfer to receiver for use.
- Mark waypoints in the field, transfer to computer for storage.
- Inexpensive, shareware programs are suitable.
- Transfers waypoints, routes and track log.

Computer-based Maps
- Map database is the most important aspect of the software.
- Provides maps on your computer screen (roads or topographical).
- Trip planning tools.
- Transfers waypoints (not maps) to receiver for use in the field.
- Transfers data marked in field to computer for display on map.
- Available from multiple vendors.
- Transfers waypoints, routes and track log.

Downloadable Maps
- Transfer of map to receiver is most important aspect of the software.
- Map displayed on computer, then transferred to receiver to display on receiver Map screen.
- Trip planning tools.
- Waypoints transfer to and from receiver.
- Waypoint management tools.

- Available only from receiver manufacturer.
- Transfers waypoints, routes, track log and actual map data.

Your use of the receiver will determine which class of program will meet your needs, but at a bare minimum all users will find a non-map program to be useful and highly affordable.

Using a receiver in conjunction with a computer requires the receiver to be capable of interfacing with the computer. Fortunately, most receivers are. However, be sure the receiver can connect to a computer before buying it. Newer models connect to the computer through a USB port, while older models use a serial port. Transferring map data through a serial port takes a long time. Connecting a receiver to a computer is simple and using the programs is easy. Using a computer with your receiver provides easier storage of waypoints and makes using the receiver more fun.

Non-map Programs

If you currently are not using a computer with your GPS receiver or you do not want to spend the money to get a map-based program, you should acquire and learn to use a GPS waypoint management program. Shareware waypoint management programs like GPS Utility, MacGPS and EasyGPS are very affordable. At a bare minimum, they make entering waypoint coordinates from a paper map much easier than typing them directly into the receiver.

Non-map waypoint programs are capable of:

- Entering waypoint names, coordinates and comments
- Forming routes
- Viewing waypoint, route and track log data on the screen
- Storing waypoint, route and track log information on a computer hard drive

Editing or entering a waypoint on the computer. ©GPS Utility Limited. Reproduced with permission.

larry_data.txt - Routes					UTM/UPS	NAD27 CONUS
Rte 03 (9)				9917841		
--- HOME-SLTLKCTY						
00	HOME		USER	--		
01	Barstow		PNT	--	1385108	
02	Las Vegas		PNT	--	1581799	
03	I-15		MINT	--	3286039	
04	I-15		MLBL	--	1681690	
05	I-15		MINT	--	1681690	
06	I-15		MINT	--	0	
07	I-15		MINT	--	0	
08	Salt Lake City		PNT	--	301515	
09						

Editing or displaying a route.
©GPS Utility Limited.
Reproduced with permission.

- Transferring waypoint, route and track log information to and from a GPS receiver

Although non-map programs display waypoints, routes and track logs on the computer screen, with their positions relative to one another, the background is blank unless the user provides a pre-calibrated map.

			UTM/UPS	NAD27 CONUS
ID	Class	CC	Coordinate	
B-LAKE	USER	--	11U 360100 5900800	
ELM-PROJ	USER	--	15T 620144 4958309	
010	USER	--	15T 626868 4976409	
GRMPHX	USER	--	12S 412013 3688091	
EAU CLAIRE	PNT	--	15T 619286 4964009	
GARMIN	USER	--	15S 343915 4302072	
FRYS	USER	--	11S 489311 3629326	
GRMEUR	USER	--	30U 607666 5648935	
GRMTWN	USER	--	51R 362926 2772402	
Road Town	PNT	--	20Q 328662 2037069	
I-15	MINT	--	12T 444807 4441912	
Salt Lake City	PNT	--	12T 421466 4514381	
Tempe	PNT	--	12S 413433 3694427	
Cave Creek	PNT	--	12S 410567 3743182	
I-10	MINT	--	12S 348870 3699789	
Hwy 85	MINT	--	12S 341371 3646904	
I-15	MLBL	--	12S 317254 4174371	
Las Vegas	PNT	--	11S 659683 4007998	

List of waypoints displayed in GPS Utility. ©GPS Utility Limited. Reproduced with permission.

Non-map programs are best used to:

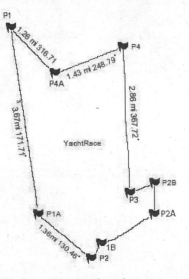

Waypoints shown in a non-map waypoint program.

- Maintain waypoint, route and track log data.
- Manipulate and enter data associated with way-points, routes and track logs using the computer keyboard and screen.
- Provide long-term storage. Waypoints may be maintained on the computer and transferred to the receiver when needed.
- Maintain receiver-independent data. Non-map programs can be used to transfer data to and from different receivers. You can transfer your saved waypoints to a new receiver or load waypoints from any receiver. If you work in a group like search and rescue, where several different types of receivers are used, the non-map program can be the medium through which data is transferred from a receiver of one manufacturer to another.

Computer-Based Map Programs

The next step up in the hierarchy of computer GPS programs are programs that provide computer-based maps in addition to all the data management capabilities of non-map programs.

The most important criterion when selecting a computer-based map program is the quality of the maps. You must first determine the type of maps you need, whether street or topographical. Then you must find a program with maps of sufficient accuracy and with up-to-date data. All computer-based map programs provide the means to manipulate and store waypoint, route and track log data, thereby making it unnecessary to use a secondary non-map program to manage your data.

Eventually street atlas and topographical map programs will all merge into a single database, but until that day, topographical map programs are provided by companies like Maptech® (Terrain Navigator, Terrain Navigator Pro), National Geographic (Topo!) and DeLorme (Topo USA) in addition to receiver manufacturer's programs such as Garmin's Topo and Topo 24k, Magellan's MapSend Topo and Lowrance's MapCreate™

USA Topo. Available road atlas programs are DeLorme's Street Atlas, Garmin's City Navigator series, Magellan's MapSend Direct Route and Lowrance's MapCreate™ USA.

Computer-based map programs also act as map databases and provide the ability to search by name or address. Most computer-based map programs offer some, if not all, of the capabilities offered by a non-map program. Keep in mind, however, that once you have marked waypoints or planned a route on the computer, only the waypoints and the routes transfer to the receiver. The map information does not transfer.

Coordinate Identification

Finding the coordinate of a location is as simple as placing the cursor on the map program and reading the coordinate. Databases of topographical maps also provide the altitude. Virtually all map databases provide coordinates using the latitude/longitude grid system, but some allow the user to select other grids such as UTM or MGRS. It does not really matter which grid is used, because you can transfer waypoints to the

Coordinates at the cursor, the circle with a cross, are automatically displayed.
© DeLorme.
Reproduced with permission.

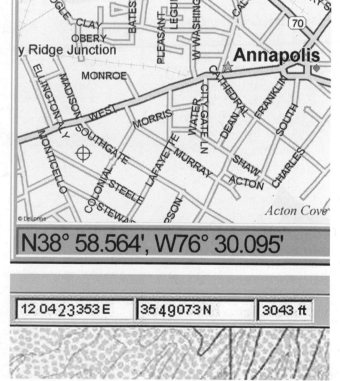

Topographical map programs provide altitude along with the coordinate.
© Maptech.
Reproduced with permission.

receiver in one format, then switch to the preferred grid and the receiver will make all the conversions.

A lot of time is spent in this book showing how to measure coordinates from a map. An electronic map makes all the manual labor unnecessary. However, do not get rid of your map rulers quite yet or forget everything you learned in the previous chapters, because unless you plan on lugging your portable computer with you on your next hike, you may still need manual techniques in the field.

A good strategy is to use the computer and the map database program to provide the coordinates of waypoints that can be identified before the trip, and carry a map with an appropriate grid into the field for everything else.

Search Engine

A search engine accepts keywords, ZIP codes, area codes, coordinates, etc., and finds locations or objects that match the specification. The matches are either highlighted or shown on a list. Selecting an object from the match list causes the program to display the object's location on the screen. Search capability is important because map databases are usually very large and cover a lot of area. Instead of panning around the map trying to find a location, simply type in anything you remember about it and let the computer do the rest. If you are trying to find Lake Hiawatha, you would type in "Hiawatha" and the program would list all potential locations:

Hiawatha Creek

Hiawatha Falls

Searching a topographical map program for all locations with "knob" in their name. © Maptech. Reproduced with permission.

Searching for 150 Elm in Castle Rock, Colorado. © Garmin. Reproduced with permission.

Hiawatha Gulch

Hiawatha Mountain

Hiawatha Trail

Lower Hiawatha Lake

Mount Hiawatha

Upper Hiawatha Lake

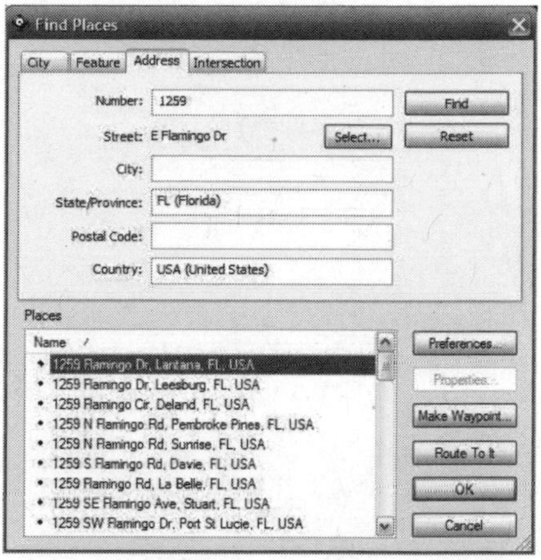

Searching for 1259 Flamingo Drive in Florida.
© Garmin. Reproduced with permission.

The computer did not find Lake Hiawatha, so you have to look at both Upper and Lower Hiawatha Lake to see if they are what you really want.

Searching works on user-specified data also. If you were setting up a cellular telephone network, you could use your GPS receiver to accurately position each transmission tower in the field. Back in the office, you would transfer all the locations to the map using a name convention to indicate the tower's kilowatt (kW) output as shown below.

1 kW Towers: T0001_1, T0002_1, T0003_1

2 kW Towers: T0001_2, T0002_2, T0003_2, T0004_2

5 kW Towers: T0001_5, T0002_5

The search engine can easily distinguish among the different types of towers, making it easy to instantly locate towers of a given power output on the map.

Marking Locations

Marking a location is another easy way to find a specific place in the future. You can mark and label any location in the map database, such as campsites, river rapids, bird sighting areas, telephone poles, old growth timber stands or anything else. Displaying a marked location on the screen is as simple as selecting its name from a list with the mouse.

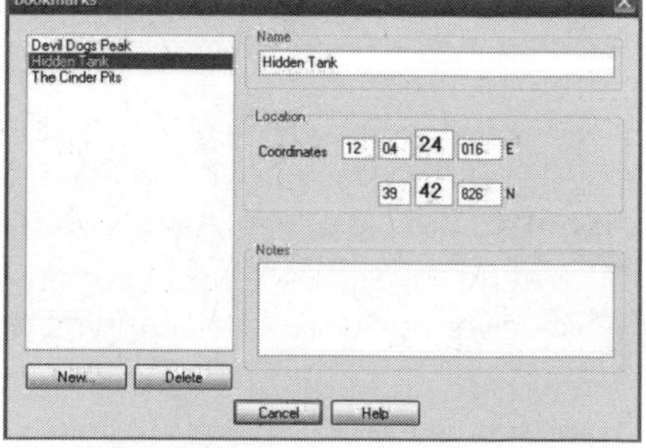

Marking the location of Hidden Tank.
© Maptech. Reproduced with permission.

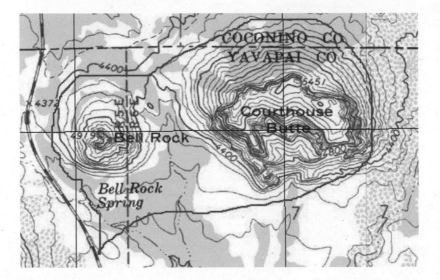

Routes are easily marked for transfer to a receiver.
© Maptech. Reproduced with permission.

Marking Routes

Much like a route on a GPS receiver, lists of successive locations may be organized in the map database to show a start, intermediate points and an end. The routes, as formed on the computer, may be transferred to the receiver to guide you in the field. Most programs will provide distance and bearing between each waypoint and total distance of the route, just like a GPS receiver.

A computer makes it possible to store hundreds of routes. When combined with a GPS receiver, the computer overcomes the receiver's limited memory and annotation capabilities. The map database program can become the route management tool for personal or business use. You can store the routes of every hike you have ever made on the computer to maintain a lifetime record and to be able to return to any location whenever you like. A uniform-laundry business could store its pickup routes on the database and transfer them to the GPS receivers in the vehicles of new, replacement or temporary drivers.

Distance, Bearing and Area Calculations

Calculating the distance and bearing between two points is easy with a GPS receiver, and of course, mapping programs can do much more. A map program can find the length of any path no matter how much it twists or turns. You can draw the path you intend on taking and find the distance before you leave, or you can record waypoints in the field, then transfer them to the computer afterwards to find out how far you went.

Area calculation is a powerful tool that may not be used by most outdoor enthusiasts, but invaluable to professionals and some specialized outdoor enthusiasts such as treasure hunters. Treasure hunters may mark areas of interest, calculate the area and file a mining claim based on the area. Professional firefighters or news reporters may calculate the area of a forest fire by entering some coordinates from the edge of the fire. The size of oil spills, water coverage, mountain acreage or a search area can all be easily measured by taking a few coordinates from the area and feeding them to the computer.

The distance from start to end is 2 miles, 4661 feet. An area of 272 acres is enclosed in the route.
© Maptech. Reproduced with permission.

Altitude Profiling

One of the best features of topographical map databases is altitude profiling. After you draw a route on the map, the computer instantly produces a profile showing the changes in altitude along the route. Try profiling a trail on a paper map and you will see the power of this feature. Profiling allows you to see in advance which sections of the trail will be challenging. Some map programs can only generate a two-dimensional profile of a path, and these are useful to professionals who design trails, plan evacuation routes, etc. Other programs can provide 3D relief pictures. Guidebook writers can use a GPS receiver in the field to accurately mark the trail, then transfer the waypoints to the map database to calculate its length and profile. Some map programs will show named points along the trail on the profile, which correlates the profile to places that are easily identifiable in the field.

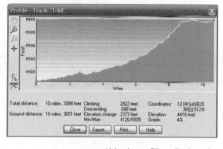

Altitude profiling displays the vertical change along the planned trail.
© Maptech. Reproduced with permission

Aerial Photos

While not available on all map programs, correlating aerial photos to a map is a very useful feature that can enhance your trip planning. By combining the USGS Topo maps with recent aerial photos of the area, you can better see the features of the terrain, as well as the condition of

An aerial photo with an overlay of a GPS track showing a mountain bike trail.
© MapTech. Reproduced with permission.

man-made features such as trails. When it comes to route planning, the ability to see the trail from an aerial photo helps you create the easiest route for your trip. Satellite photographs are available through Google Earth, Microsoft Virtual Earth, TerraServer USA (subscription service), and MapTech (part of Terrain Navigator Pro). The USGS provides some aerial photos but they are not as recent as other services.

Unfortunately, maps of backcountry areas usually do not have as high a resolution as urban areas, so you may not be able to accurately identify important features there.

Printing

There is no need to buy separate paper maps in addition to a map database program, because maps can easily be printed from the database for use in the field. The user specifies the area to be printed, the scale, the amount of detail, and whether a grid is to appear on the print copy. User information added to the database can also appear on the printed version, so when you need to see waypoints you have entered, the information is superimposed on the map.

park 1

Currently, map database programs can put a grid on the map. However, many times, the grid is not directly usable for finding coordinates without additional preparation. All the techniques taught in the previous chapters will help you deal with whatever grid the map database program produces. If you always print maps at the same scale, you can even make a ruler to quickly find coordinates. Fortunately, map programs are improving daily and someday the grid on the map from your database program will be useful just as it is printed.

Downloadable Maps

Electronic maps may be transferred to a GPS receiver in one of two ways. Many receiver manufacturers sell map database programs that can transfer map data to the receiver over a USB connection. Other manufacturers sell memory cards (e.g., SD cards) preloaded with map information. Once the memory card is inserted into the receiver, the receiver can display the maps. No connection to a computer is required.

A map program that transfers the map data from the computer to the receiver has all the capabilities listed above for non-map and computer-based map programs, with the possible exception of aerial photos. Downloadable map programs display the maps, waypoints, routes, etc. on the computer screen, then permit the user to transfer the map and the other information to the GPS receiver for display on the receiver's Map screen. The map shown on the receiver screen is exactly the same map seen on the computer. Downloadable maps make paper maps obsolete, as long as your receiver is functioning properly. As you travel, you can watch your progress on the downloaded map and see all roads or mountains go by on the receiver's screen.

Obviously, because of the difference in screen size, the computer will show more area with greater detail than the receiver will; however, any street or geographical feature that can be seen on the computer can be seen on the receiver if zoomed in properly. The important point with downloadable maps is that the information available on the computer can be downloaded to the receiver and displayed in the field. A paper map is still useful in the field to provide a view of a larger area.

Generally, maps capable of being downloaded to a receiver are available only from the receiver manufacturer and they generally cost more than the computer-based map programs described earlier. If your use of a GPS receiver requires downloadable maps, and especially if you plan on traveling in foreign countries, check with the manufacturer before you purchase to ensure that the maps you need are available. Another important thing to note is the detail of the map to ensure that it is suitable for navigation purposes. Topographic maps at the 1:100,000 scale, which is a common scale for downloadable maps, can provide some insight as to your general position in relation to the terrain, but they are unsuitable for navigation and cannot in any way replace a paper map. Downloadable maps at the 1:24,000 scale are suitable for navigation, as they have the same detail as paper maps.

10 Using a Computer and Topographical Maps

The previous chapters illustrated how to take coordinates from a map, program them into a receiver and use them on a trip. A computer can remove the drudgery of measuring coordinates from a map and the laborious process of typing them into a receiver. This chapter illustrates how computer-based maps can be used to plan a hiking trip.

This example uses a topographical map program called Terrain Navigator Pro, produced by Maptech®, which includes access to aerial photographs synchronized to the USGS topographical maps. Other programs such as National Geographic Topo! or DeLorme Topo USA have similar features, except possibly the aerial photographs. Aerial photos show the condition of the area and any changes since the map was last updated. The software may be used to plan trips and transfer waypoints and routes to the receiver, in addition to uploading and displaying any waypoints marked in the field.

You have visited Chiricahua National Monument in southeastern Arizona several times over the years. You find the strange landscape fascinating. You have a trip planned for the area, and this time you will climb to the top of Chiricahua Peak, spend the night there, then descend the next day. You immediately install your new topographical map software and search for the word "Chiricahua."

The Place Finder window searches the maps and finds all the places that have the word "Chiricahua" in them. You are amazed to see a place called Chiricahua Natural Bridge. You have never heard of it. You double click on the name and a map instantly appears showing the location of the natural bridge to be close to the monu-

The Place Finder window reports back ten locations with Chiricahua in their name.

ment headquarters. You select the Marker tool and place a marker near the bridge. You access the Edit Markers menu and change the automatically generated waypoint name "Natural Bridge."

The Edit Markers window also displays the coordinate of the waypoint as well as the name that will be displayed on the GPS receiver (e.g., NtrBr)

Renaming the waypoint marker. Its coordinate is also shown.

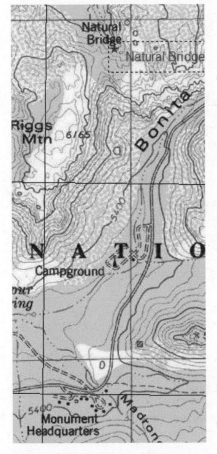

Monument Headquarters at the bottom, Natural Bridge at the top.

when the waypoint is transferred. Getting the coordinate of the natural bridge was as easy as clicking the computer mouse. You did not have to use a ruler or eyeball it. It was so easy to get the coordinate, you wonder why you waited so long to get map software.

The contour lines between the road near the monument headquarters and the natural bridge show the elevation changes from approximately 5400 ft. to about 5960 ft., but the elevation lines are close. You switch to three-dimensional (3D) mode to visualize the change in elevation. The 3D view shows a cliff between the bridge and the road. Rotating the 3D view reveals that the bridge is isolated and difficult to get to. The map does not show any trails leading to the bridge from any direction. You put it on your list of places to explore.

You return to the place finder window and double click on "Chiricahua Peak." Instantly, the

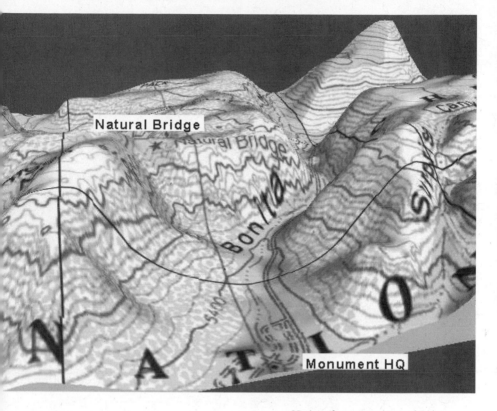

Natural Bridge

Monument HQ

3D view of area around Natural Bridge.

peak appears on your computer screen. You zoom out and look for possible trails leading from the peak. Rather quickly you determine the best path is to leave from the Herb Martyr Campground, hike an established trail to Cima Park, then on to Chiricahua Peak. There, you will spend the night and return the next day by way of Snowshed Peak and Pine Park.

Using the mouse, you draw a line to mark your route along a trail shown on the map; however, some USGS topographical maps have not been updated for at least a couple of decades, so it is possible the trail no longer exists. This is where purchasing a map program that accesses aerial photographs is advantageous. You zoom in on the area you will be hiking, switch to aerial photo mode and search along the trail you marked. You can see from the photos that the road to the campground has changed, but you can also distinctly see parts of the trail. At the locations where you can see the trail, it looks well established. The trail is most visible where it cuts through vegetation. Enough of it is visible that you believe your hiking plan is feasible, but you also know that even

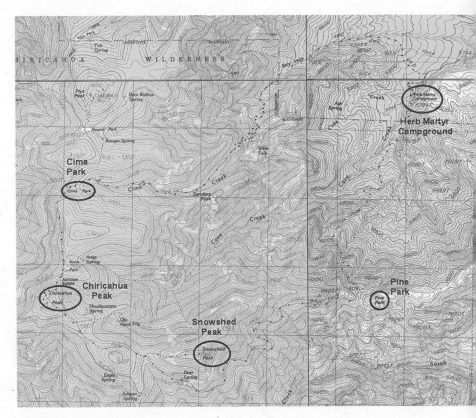

Map of proposed trip, shown as a dotted line.

Aerial photos allow you to see the existing trail.

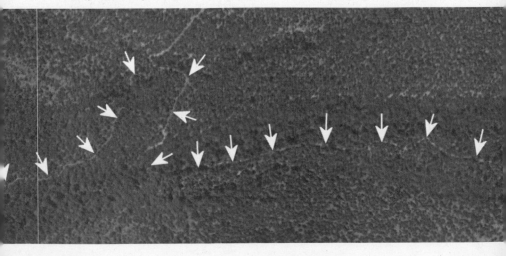

before arriving, the return trail may be hard to find. This technology is so powerful that you're already wondering when you will be able to look at live satellite images of a planned hiking area right from your desk.

Your proposed route looks good, so you want to change the line on the computer screen into a route you can use in your GPS receiver. You activate the Create Route window. After you type in the route name, you are offered several options. You select the option where the route generated is within 100 ft. (30.5 m) of the drawn line. An information box pops up to report the results of the conversion. The greatest deviation from the drawn line is 98 ft. (29.9 m) and there are 86 waypoints in the route. The waypoints are automatically named Wpt1, Wpt2 and so forth.

Converting to a route.

You wonder how close the route really is to the line you drew over the marked trail. You zoom in to see that the route with its waypoints closely corresponds to the trail. You activate the route waypoint list and select "Wpt9." The waypoint coordinate is displayed on the map and is measured to the meter. Getting the waypoint coordinate did not require a ruler, estimating or any typing on either a computer or a receiver. Map software makes getting waypoint coordinates simple.

Displaying a waypoint from a route.

The map program provides other tools that give you invaluable information. Selecting the route you have traced, you activate the profile view. The profile shows the distance of the entire route as being just over 13 miles (20.9 km). The profile also shows the distance from the campground to the camping place southeast of the peak (the highest point on the profile) to be just about 6 mi. (9.7 km), but the vertical ascent is 5013 ft. (1528 m). For you, that much change in elevation in a single day is a hard hike. You search the map for a nice camping spot halfway to the summit. You are unable to find a suitable location. The best camping areas are all around 9000 ft. (2743 m), which is almost as high as the peak. You decide to modify your plan to start the hike earlier in the day and to stay an extra day on the peak to rest, explore and enjoy yourself to make up for all the hard work of getting there.

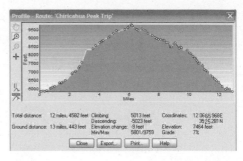

Profile of the hiking trip.

3D view of the climb to the peak.

Before you go any further, you print a copy of the map and the intended route so you can have it with you on the trail. While you are printing, you notice heavy lines on the map. You know that the thinner lines are UTM grid lines, but you do not know what the heavier lines mean. You select the information cursor and select the lines. You learn that the heavy lines represent the boundaries of the USGS 1:24,000 paper maps. The computer program abuts the maps so they seamlessly cover the entire state. Your route uses four USGS maps: Chiricahua Peak, Rustler Park, Portal and Portal Peak

You are about done planning your trip. The planning went faster than ever before and you know a lot about the trail and the nature of the hike even before arriving. You need to transfer the waypoints to your receiver, so you activate the screen that allows you to select the receiver type. You select your receiver and the port to which the receiver is connected. You start the transfer and with a few clicks of your mouse the entire route is on your receiver.

Because you have a newer receiver, it accepts the route even though there are 86 waypoints. Older receivers may limit the number of waypoints per route. Should this problem occur, there are ways around it. First you may regenerate the route and limit the number of waypoints created to 50. Reducing the number of waypoints will reduce the accuracy of the route in relation to the trail. You may also delete waypoints from parts of the route by hand, in areas where deletion will not create inaccuracies. Another option is to break the trip up into two routes. By creating a separate route for the two different legs of the trip, each route can have 50 waypoints.

On the receiver you go to the route list to look at the route you just transferred. A scan of the waypoints shows that they are all there and the total length of the route is 12.9 miles. You select one of the waypoints. The coordinate checks out when you eyeball its position on the map. It took hardly any work on your part to get accurate coordinates for an entire trip. If you were not so emotionally attached to your old paper maps, they would end up in the garbage tomorrow.

The day of the trip arrives. Early in the morning, you fire up your receiver and start up the hill. It is every bit as hard a climb as you thought it would be, but the views at the top and camping for two days in total peace make it worth the effort. While exploring the area around the peak, you mark several waypoints.

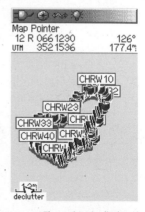

The entire trip displayed on the receiver's screen.

You know you will be able to transfer them from your receiver to the map program when you get home. You have stopped wondering why you did not buy the computer map program before and are just glad you did.

Using Electronic Topographical Maps on Your Receiver

Now that you are comfortable using topo programs on your computer, you want to make the process even easier: by taking the topo map with you on your GPS receiver. The ability to download topo maps into a receiver and use them for navigation has been available for some time, but had been limited in the past by two important factors: storage space, and detail. Electronic maps require large memories to store them. In older-model receivers, memory capacity was so small that it was filled simply storing the waypoints, tracks and routes you created during your travels. With the advent of SD and microSD memory cards, gigabytes of memory are now available for your receiver. Furthermore, if one memory card cannot hold all your electronic maps, cards can easily be switched, allowing for the storage of new maps and more information. These cards are different from the standard internal memory of your receiver in that they can be removed and exchanged for another card in a matter of seconds.

Until recently, the detail of the electronic topographical maps was another limiting factor. Many electronic topo maps use USGS map data at a 1:100,000 scale. A 1:100,000 scale map may provide some useful information about the terrain, but it does not provide sufficient detail for navigation. Newer electronic topo maps use the 1:24,000 scale, which is the same detail provided by the paper and computer topo map programs, and therefore are excellent for navigation. Combining the electronic topo map with your receiver and route-planning tools can be a valuable asset. While the electronic map cannot fully replace the big-picture view of a paper topo map, it can put in your hand a much more informative picture of your route using the receiver alone.

Electronic topo maps are available through the manufacturer of your receiver. Unfortunately, electronic topo maps at the 1:24,000 scale are available for only a tiny portion of the U.S. While they may vary in detail from area to area, many provide information as to trailheads, camping areas and other points of interest. Some topo map programs have auto-routing capabilities for hiking trails, thereby making your planning easier and more efficient.

Because a topo map of the entire United States would require more memory than is available, the map programs allow you to select an area of interest, called a region, which fits on the internal memory of your

GPS receiver. The larger your internal memory, the more regions you can store. If you travel a lot and wish to have electronic topo maps for numerous regions, you can store different regions on several cards or buy cards with greater data storage capacity.

After you transfer and store the regions in the internal memory of your receiver, the region you desire to view can be loaded into the active memory. Routes, tracks, waypoints and other information will be displayed over the receiver's base map.

Preprogrammed Data Cards

Another way to get topo maps and easily use them is to buy preprogrammed memory cards that can be inserted into your receiver. Purchasing preprogrammed cards eliminates the need to acquire a program that runs on your computer. Newer receivers have both internal memory and a slot for inserting memory cards. You can buy cards preloaded with electronic topo maps from various regions, but not all regions have maps at the 1:24,000 scale. Because memory cards are small compared to the amount of data needed for 1:24,000 scale maps, available areas are typically broken up into separate regions to make purchase convenient. For example, one manufacturer divides the U.S. into 11 different regions, with some overlap between them. Users should need to buy only one or two cards to cover the areas they are most likely to visit.

As mentioned earlier, electronic maps at the 1:24,000 scale are not available for all areas of the U.S. If you need 1:24,000 maps for areas not covered by the downloadable ones, you will need to purchase either a paper map or a computer topo map program that has the maps of the area you want.

Navigating with Downloadable Topo Maps

For some time now you have lived near but have never visited the Grand Canyon. After arranging your work schedule, you plan to take some time off and hike along the south rim of the canyon. You have been told that the Tonto Trail along the south rim is a nice hike with beautiful views of the canyon. As downloadable topo maps can only be obtained from the manufacturer of the receiver, you purchase the appropriate software that corresponds to the receiver you use. Since you are accustomed to using paper topographical maps, you want an electronic map that will provide the same detail as the 1:24,000 scale paper maps. While 1:100,000 scale electronic maps are available from various manufacturers for the entire U.S., 1:24,000 scale maps are currently available only through Garmin as part of the National Parks U.S. Topo 24k package. No electronic maps at the 1:24,000 scale are available for any entire state, only for selected national parks.

The National Parks package divides the U.S. into three regions: West, Central and East. Each region has 1:24,000 scale maps for selected national parks in that region. You live on the west coast, so you select the National Parks West package, which includes the states AZ, CA, CO, HI, ID, MT, NM, NV, OR, UT, WA and WY. You check to make sure it includes some of the national parks you commonly visit, and are pleased to see the Grand Canyon and Yosemite as well as several of your favorite parks in southern Utah. After installing the software, you pull up the Place Finder to quickly locate the Tonto Trail.

Like any other computer topo map program, the National Parks package permits you to mark waypoints, quickly find coordinates and form routes. It also has all the other features you would expect in such a program. However, the National Parks package also provides auto-routing along known trails. Just like a GPS receiver in a car, auto-routing permits the user to select a start point and an end point, and the computer program will form a route that follows established trails. You load the program into your computer and begin thinking about where you want to go.

As you ponder, you realize that the only thing you know about the Tonto Trail in the Grand Canyon is what you have heard from friends. You do not even know the location of the trailhead. You select the Tonto Trail from the Place Finder window. The trail appears on the computer screen. Using the mouse, you scroll along the trail and in seconds you find the trailhead near a 4WD road that leads to the main highway. As you scroll along the trail, you realize that it is actually quite long. You know that forming a route along the trail by hand could take some time, so you take advantage of the auto-routing capability. With one click of the mouse, you mark the start of the 4WD road as the starting point of your route. Two clicks later you have formed an 11.8 mi. (19 km) route from the 4WD road along the entire stretch of the Hance Trail to the Tonto Trail and along a portion of the Tonto Trail to an intersection with another trail. You smile to yourself. Auto-routing is great for cars along roads, but it is even better along trails in the backcountry.

Using the Place Finder window to find the Tonto Trail.

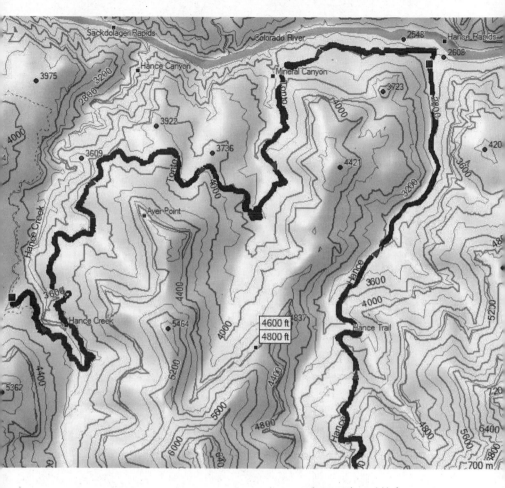

Auto-routing quickly forms routes along known trails with only a few clicks.

Pleased at how easy this has been so far, you continue using auto-routing to form the route along the trail. After routing another 30 miles, you run into an intersection with another trail. You place the computer's mouse pointer over the trail. The name of the trail, Kaibab Trail, appears on the screen. By looking at the contour lines, you can see that the trail switchbacks down the canyon until it intersects with the Bright Angel Trail, which follows the Colorado River for some distance before climbing out of the canyon and intersecting with the Tonto Trail. Having never hiked down the Grand Canyon, you think this could be an exciting detour, albeit strenuous, judging from the contour lines.

Looking at the area, you notice there is a bridge at the bottom of the canyon that crosses the Colorado River, allowing access to a campsite.

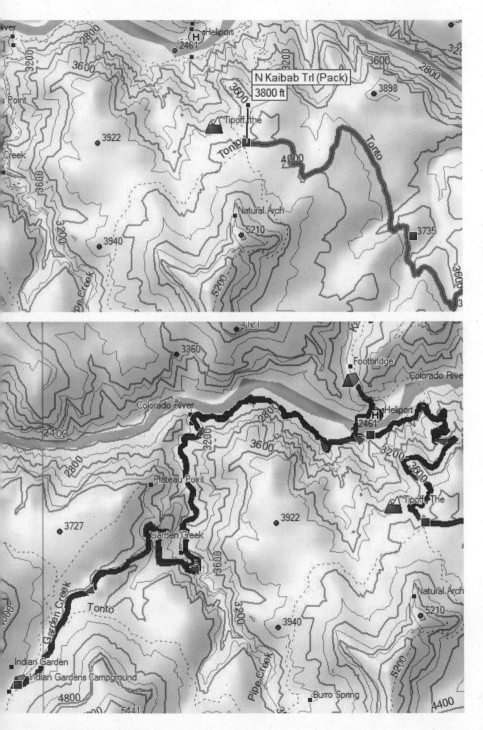

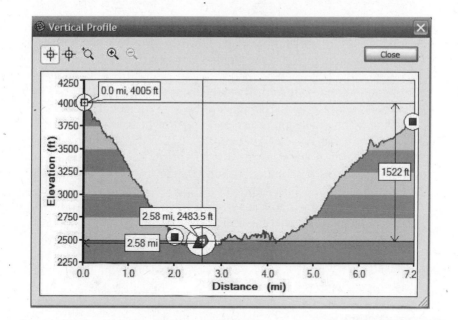

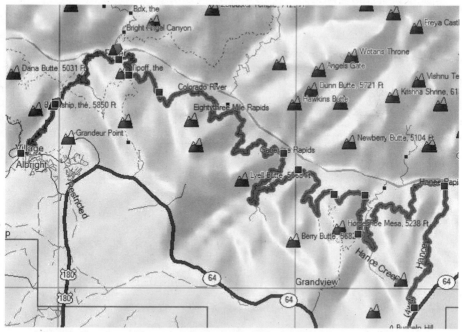

opposite top: Planning a detour along the North Kaibab Trail.
bottom: With two clicks you plan where to camp and your trail.

this page top: Examining the strenuous profile of the trail.
bottom: The proposed route. Auto-routing made route formation simple.

You quickly click and create a route for a detour from the Tonto Trail down to the Colorado and back up, allowing for a stop in the Bright Angel campground as well as the Indian Gardens campground.

You know the trail will be difficult, so you look at the profile for the little detour you have created. The section of the trail you selected for the detour is only 7.2 miles from the point where you depart from Tonto Trail to the point where you return to Tonto Trail at the Indian Gardens campground. While you can get a general idea of the elevation change from just looking at the scales at the side of the profile, you want to know the exact distance and elevation change from where you leave the Tonto Trail to the Bright Angel campground. Using the Marker tool, you discover that in the 2.58 miles you will travel, you will descend 1522 feet. Because you can continue along the Tonto Trail for the entire rim, you decide to stay on Bright Angel Trail until you arrive at the Grand Canyon Park entrance, where you will arrange for a vehicle to pick you up. Reviewing the route you have formed, you see that the total length of the proposed trip is 53.6 miles (86.3 km). It took only a few minutes and a few clicks of the mouse to form the route, because auto-routing did most of the work.

A more feasible route.

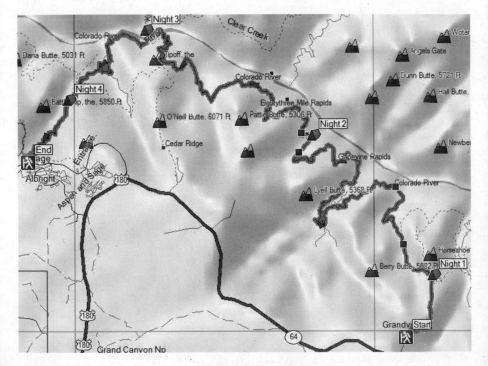

You realize that in your exuberance to auto-route you may have bitten off more than you can chew. You decide to see if you can access the Tonto Trail through a different point than the Hance Trail. You scroll the map across the screen to follow the freeway. You see another road and a trail called the Grandview Trail leading to the Tonto Trail at the intersection you noticed previously. You can see that entering onto the Grandview Trail will reduce the total length of the hike and possibly leave you with enough energy to negotiate the Bright Angel Trail. Because auto-routing makes it so easy to form routes, you delete the first route and use the computer mouse to form a new route that includes the Grandview Trail.

Now that you have a better idea of the trails you will follow, you plan your excursion with more detail. Since you plan to be out for four nights, you need to find two additional campsites. Using the topo map on the computer screen, you find two sites that are well situated on large plateaus. You then create a route to pass through the four selected campsites. When you are done you have a trail that is 34.1 miles (54.9 km) in length.

Using the Profiling tool, you can examine each leg of the journey. You create a table showing the distance and change in altitude of each leg. Knowing the trip will be strenuous no matter how you approach it, you feel confident that the planned route will be manageable by you and your friends. Since you will be starting later in the afternoon on the first day, you keep the distance of the first leg short, although the descent will be significant. You plan to cover a lot of territory on days two and three, leaving time on days four and five to slowly make your way out of the canyon.

Day	Distance	Descent/Ascent
One	2.3 mi. (3.7 km)	–2395 ft.
Two	12.8 mi. (20.6 km)	–1242 ft.
Three	9.7 mi. (15.6 km)	–1201 ft.
Four	4.7 mi. (7.6 km)	1351 ft.
Five	4.6 mi. (7.4 km)	2991 ft.

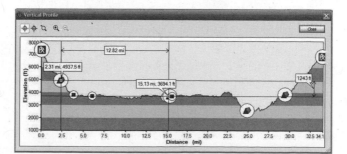

The profile for Day Two of your backpacking adventure.

You are astounded at how quickly you were able to able to create a route, plan campsites and analyze your week-long backpacking trip. You excitedly connect your GPS receiver to the USB port of your computer so you can transfer the topo

Maps(4)	Waypoints(2)	Routes(1)	Tracks

Name /	Area	Size
Cape Royal, AZ	Grand Cany...	832 KB
Grand Canyon, AZ	Grand Cany...	638 KB
Grandview Point, AZ	Grand Cany...	409 KB
Phantom Ranch, AZ	Grand Cany...	699 KB

Four 1:24k topo maps easily fit in the receiver's memory.

maps, the route and the waypoint information to your receiver. You want to transfer the correct electronic topo maps to the receiver, so you select the route and choose the option "Select Maps Around Route." The computer program identifies the USGS topo maps that include the route. The route covers a portion of four different maps. The computer prepares to transfer the four electronic maps and the route data. The data occupies 2.55 MB of memory. Your receiver has sufficient memory, but before making the transfer, you specify that the route calculation data also be transferred, because your receiver is capable of auto-routing. Two more clicks of your mouse and the data is successively transferred from the computer to your GPS receiver.

The route after transfer does not follow the trails.

When you display the transferred route on the receiver's Map screen, you instantly realize that something is horribly wrong. Instead of displaying a route that follows the trail, as on the computer, the receiver shows straight lines between the six points you marked on the maps. You realize the auto-routing capability is not enabled on the receiver. Like the computer program you used to originally form the route, your GPS receiver, if capable of auto-routing, can also form routes along established trails. You enable auto-routing and much to your relief the route conforms to the established trails within seconds. The route on the receiver now looks exactly like the route formed on the computer.

Enabling auto-routing on the receiver.

A receiver with auto-routing capabilities is a powerful tool. Unlike the previous example above, which required over 80 waypoints to create a much shorter route, auto-routing enables you to create a route that precisely follows a trail by indicating, in this case, as few as six waypoints. Once on the trail, if you decide to take another trail, the auto-routing capability of the receiver will form a new route to guide you along the newly selected trail. With receiver auto-routing, routes are formed in the field with a few clicks of a button.

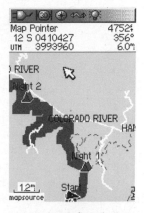

Auto-routing confirms the route to the selected trails.

You look at the map on the receiver's screen and are amazed at the details shown. The contour lines are just like those shown on a paper 1:24,000 scale USGS topographical map. As you zoom in and out on the Map screen, the amount of detail automatically adjusts to provide information without cluttering the screen. You move the cursor (e.g., white arrow) around the receiver's Map screen and note that placing the cursor over a location provides the coordinates and elevation of the location. The receiver provides all the capability of the computer, except on a screen small enough to be carried into the field.

Contour lines shown on the receiver's Map screen provide the same amount of detail as a paper map.

Zooming in and out can provide more detail or less. The receiver cursor provides a location's coordinate and elevation.

You select the Elevation Profile tool on the receiver. The screen provides a profile of the elevation along the route just like the computer did.

The National Parks package does not include aerial photos like Terrain Navigator Pro, discussed above, but waypoints may be uploaded into Google Earth™ and displayed on a background of satellite photos. Unfortunately, Google Earth does not perform the auto-routing and will show only the straight-line route between waypoints. But Google Earth does display trail information, so you may get some insight into the terrain before you arrive.

Topographic maps with auto-routing can be a tool that makes planning a trip fast, easy and convenient. These advanced features also give you more flexibility when it comes to changing your route or destination, because with only a few clicks of a button the receiver can calculate a new route that follows a different established trail. GPS receivers get better every year.

The receiver provides an elevation profile.

11 GPS on the Road

GPS receivers for automobiles are among the most advanced receivers on the market. They combine electronic maps, automatic route generation and GPS position calculations. GPS receivers designed for cars can be significantly more expensive than the typical hand-held receivers presented in earlier chapters, so it is helpful to understand what they can do that less expensive receivers cannot.

The two most important features that distinguish automotive receivers from other receivers are:

- Automatic route generation (auto-routing)
- Turn-by-turn instructions

Auto-routing means the receiver is capable of plotting a route from one point to a destination, along roads, highways or trails, without any input from the user. Auto-routing is different from routes formed using downloadable maps or routes formed by manually entering sequential waypoints.

A receiver capable of accepting downloadable maps, but incapable of automatically calculating a route, will show the downloaded map, but it will not display a route that follows the roads. Assume you download a street map into your GPS receiver. Assume also that you use the computer to calculate a route between two points (e.g., a starting point and a destination). The route calculated by the computer and displayed on the computer screen will follow the roads and highways of the map. The computer may even provide turn-by-turn instructions. However, if you transfer the route to a GPS receiver that does not have auto-routing, the route will be displayed on the receiver screen as a straight line between the start and the destination. The route, on the receiver, will not follow the roads.

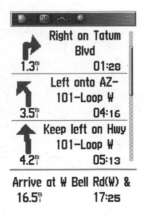

Turn-by-turn directions.

159

Computer generated route shown on computer screen.

	Directions/Name	Distance	Leg Length	Leg Time	Course
1.	School	0 ft			
2.	Get on N 59th Ave and drive south	0 ft	0 ft	0:00:00	270° true
3.	Turn left onto W Utopia Rd	48 ft	47 ft	0:00:02	180° true
4.	Turn left onto N 57th Ave	0.4 mi	0.3 mi	0:01:08	86° true
5.	Turn right onto W Beardsley Rd	0.8 mi	0.4 mi	0:01:23	12° true
6.	Take the Az-101-Loop E ramp to the left	1.5 mi	0.7 mi	0:01:03	84° true
7.	Keep left onto Az-101-Loop E	3.7 mi	2.2 mi	0:02:11	91° true
8.	Keep left onto Hwy 101-Loop E	10.7 mi	7.0 mi	0:06:29	88° true
9.	Take exit 41 to the right onto Shea Blvd	21.8 mi	11.0 mi	0:09:48	179° true
10.	Keep left onto Road ramp	22.2 mi	0.4 mi	0:00:46	179° true
11.	ScottsdaleHeal	22.6 mi	0.4 mi	0:00:32	180° true

Turn-by-turn instructions generated by the computer.

As seen in the GPS screenshots, the solid black triangle represents your car. The line indicated by the light track is the route between your current position and your destination. It is impossible to drive the straight route, because it does not follow the roads. Although the route was correctly calculated and displayed on the computer when it was transferred to the receiver, it became meaningless because the receiver is incapable of making the route follow the roads. As you can see, a receiver capable of auto-routing produces the same route as you had previously calculated, following the roads until it reaches the final destination. The computer screenshots show an automatically generated route between 5900 W. Utopia in Glendale, Arizona, and Scottsdale Shea Hospital in Scottsdale, Arizona. Not only is every required turn in the list, with the requisite distances, but also a description of how the turn is to be made. Explicit turn-by-turn instructions make it possible to make the trip without looking at the GPS receiver display; however, the map on the receiver display allows you to see the big picture.

Route displayed on receiver without auto-routing capability.

Most auto-routing-capable receivers permit the user to specify priorities the receiver is to use when calculating a route. When calculating a route, the receiver consults the user priorities each time it must make a choice between two possible roads.

Generally, users may specify priorities such as the following.

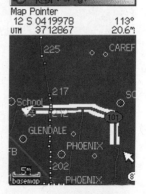

Auto-routing conforms the route to the road, on the receiver.

Form a route that provides:
- Faster travel time
- Shorter travel distance
- Most direct route

Route should avoid:
- Highways
- Toll roads
- U-turns
- Unpaved roads
- Traffic
- HOV Lanes
- Ferries

If you specify that routes should be formed to provide faster travel time, receivers will route over highways as opposed to local streets whenever the choice arises. Selecting routes that provide shorter distance traveled will result in routes that use more local streets than highways, unless the highway provides the shortest distance. Instructing the receiver to avoid various types of road features, such as toll roads and U-turns, will accomplish just that.

If you instruct the receiver to avoid traffic, the receiver cannot successfully comply unless it receives timely traffic information. More advanced receivers are equipped to receive traffic information via radio broadcast. Traffic information is provided by subscription services and is available only in select cities, so if you want routes that avoid traffic, be sure the receiver you buy can receive the traffic radio signals and that traffic information services are available for your city.

Another important aspect of auto-routing is how the receiver responds when you stray off the planned course. If you run into road construction or simply miss a turn, a receiver capable of auto-routing will determine a new route from your current position to the destination. Automatic recalculation of the route when you stray permits you to take a road you know is better than the calculated route, yet depend on the receiver when you are in unfamiliar territory. Auto-routing enables the receiver to automatically update the route regardless of how far astray you go.

Automatic recalculation of the route also helps to cope with map inaccuracies. No map, regardless of whether it is paper or electronic, is completely accurate. Roads are always changing and it is next to impossible to ensure that a map is completely up to date. Not only can new roads not appear on a map, but sometimes proposed roads yet to be built find their way onto a map. Regardless of the cause of the error, the situation will occur where the map on the receiver's screen tells you one thing, yet the road in front of you tells you something else. In such a situation, you obviously follow what you see and let the receiver recalculate the route based on the maneuvers you make.

Calculating a route requires a lot of processing power, so before you buy a receiver, see if you can test it or talk to your friends to see if calculating routes is done in a reasonable amount of time. If the receiver takes too long to determine a new route, you could miss a critical turn and be stuck on a freeway, headed the wrong direction, before the receiver can tell you otherwise. Obsolete or old map data also forces the receiver to recalculate. Newer-model GPS receivers can calculate a new route in 3–5 seconds from the time you depart from the indicated route, which is plenty fast for you to respond appropriately to the new directions.

Memory is also important. The maps used by the GPS receiver either have to be downloaded into the receiver's built-in memory or onto a memory stick. Most receivers capable of auto-routing accept memory cards, which may be bought either blank or with maps already loaded. Models that are designed specifically for use in an automobile are frequently equipped with a built-in memory system, either a hard drive or flash memory, which holds all necessary map and data such as points of interest. Large capacity built-in memory eliminates the need for memory cards or transferring maps from your computer to the receiver. The more memory available for maps, the more territory you can cover using highly detailed maps.

All mapping receivers have a base map, so even if you do not purchase more detailed downloadable maps, you can still navigate in many areas, usually larger cities. Detailed maps include points of interest such as hospitals, rest areas, restaurants, shopping, entertainment venues, emergency services, amusement parks, museums, banks and tourist attractions. You may take a number of approaches to provide your receiver with the memory necessary for all the maps you may need. You may purchase several memory cards and split the maps you need among the cards; purchase one large-capacity card capable of holding all the necessary maps; take your laptop with you to transfer maps to the receiver as needed; or purchase a GPS receiver preloaded with maps of the entire area you expect to travel in, such as all of the U.S. or Europe.

Another thing to consider is the antenna. Some GPS receivers designed for vehicles come with external antennas for placement on the roof of the car. If the receiver you choose does not include an external antenna, purchase one. The screen on automobile receivers is larger than most hand-held receiver screens, but you may at times want to take a closer look. If you have an external antenna, you can take the receiver off the dash and either look at it more closely yourself or pass it to another person in the car, even if they are in the back seat. An external antenna provides stronger satellite signals than a receiver's internal antenna. The cost of an external antenna may seem like a lot, but the flexibility and convenience it allows when using the receiver is worth the money.

Most receivers designed for automobiles can also talk to the user to provide audible driving instructions. Voice instructions are very convenient because they free you from looking at the screen each time there is a change in direction. If you are driving alone, voice instructions can make your drive much safer.

Use is Simple

Using a GPS receiver designed for an automobile is simple. First, if the receiver does not already have all the maps required, download them into the receiver or memory card. The next step depends on whether you are already in the car or are planning a trip in advance. If you are already in the car and the receiver is locked onto the satellites, select where you want to go, tell the receiver to create a route, and it will guide you to your desired destination. It is as easy as that. If you are planning a trip in advance and wish to store it in the receiver's memory for later use, you need to select a destination and store it in memory. When the day comes to make the trip you simply select the destination, and the receiver calculates a route from your present position to the destination.

Selecting where to go is easy. As mentioned above, the electronic maps have not only road information but also locations that are considered a point of interest (POI). Finding a hotel or restaurant is simple because the receiver allows you to search for POIs much like you search for waypoints on a hand-held receiver. You can either search for a specific type of place or by name. Have you ever been on a long trip in an unknown area and

Choose from restaurant types.

Locate important community centers.

Search for stores and shopping.

wondered where the closest service station, restaurant or hotel is? The receiver can list the nearest POIs and then guide you to the one you select. Destinations may also be found by address or even by intersection.

The points of interest are not all-inclusive. If the map database does not contain a specific location, it cannot lead you to it. If you have a favorite hotel or restaurant chain, you may want to ask, before you buy, if your favorite locations are part of the POI database.

Visiting your friend's cabin

Summer has finally arrived and you decide it's time to flee the city to the cool breezes of the mountains for a summer adventure. A friend kindly offers you the use of a cabin in Taos, NM. Your friend is already in Taos and the cabin does not have an address that she knows of, so you agree to meet in Taos. She also tells you to take Highway 260 out of Payson. As you leave your driveway to head up to Taos, you turn on your GPS receiver to select a destination. You choose from the cities category and type in the name Taos when prompted. A list of cities named Taos appears and you choose Taos, NM, as your destination. Five seconds later the receiver has generated a route and you are off. You follow the receiver's voice instruction until you arrive at Payson, AZ. You notice that the GPS receiver tells you to continue north on the Beeline Highway, but your friend told you to head east on Highway 260. You know the receiver's route will get you to Taos, but you decide to take Highway 260.

Select Taos, NM, from a list of cities.

You can prompt the receiver to calculate a new route using Highway 260 in one of two ways. First you can turn onto Highway 260 and let the receiver calculate a new route based on your current position and direction of movement. Unfortunately, the receiver will try many times to get you to turn back onto the Beeline Highway. You will need to ignore it until you get far enough down Highway 260 that the receiver finally calculates a new route that goes that way.

What if you want to take Hwy. 260 instead?

You may also insert a Via Point to get the receiver to route along Highway 260. A Via Point tells the receiver to route to the destination via the point you indicate. The receiver you have is a touch-screen model, so you touch Highway 260 on the map with your finger. Touching the highway causes the receiver to ask if you want to insert a Via Point. You tell the receiver to insert a Via Point on Highway 260 and the receiver calculates a new route to Taos, NM, through the Via Point indicated on Highway 260. Using the Via Point was simple and it allows you to follow the highway you want to follow but still have the receiver guide you the rest of the way.

Inserting a Via Point.

The drive continues peacefully and before you know it you are in New Mexico. As you cross the state line you enter a heavy thunderstorm. You suddenly realize that your wipers have cooked in the Arizona sun for the last two years and are completely useless against the rain. You need new wipers, now, so you turn to your receiver. You search POIs in the category of Auto Services, where you find the name "Autozone," which happens to be your favorite auto supply store. You insert "Autozone" from the list as a Via Point in your current route. The receiver then directs you to your destination via the Autozone store. After you stop and replace your wipers, you get back on the road and continue the indicated route. Eventually, you arrive in Taos and meet your friend.

Rerouted to the same destination via Hwy. 260.

Via Points, as demonstrated above, can be a valuable tool in navigating roads. You can search for places that are near your current location, your route, or your destination and insert them as stopping points in your route. The receiver will guide you to these points and then on to your destination.

Searching for the nearest Autozone.

Select "Go" for Autozone, then insert it as a Via Point.

12 PDAs, Pocket PCs, Laptops and More

GPS technology has been combined with Personal Digital Assistants (PDAs), Pocket PCs, laptop computers, radios and cellular phones. Thanks to recent advances in technology, it is increasingly possible to add GPS functionality to the electronic devices you use every day. Before we discuss some of the possibilities of integrating GPS use with commonly used electronic devices, it is useful to look at some of the advances that have made this possible.

Smaller Chipsets

By decreasing the size of the chips (e.g., integrated circuits) that enable GPS satellite signal reception and position calculation, it has become possible to add GPS functionality to many devices.

GPS-enabled cell phone.

Bluetooth

Bluetooth is a wireless communication protocol. Most wireless cell phone headsets use Bluetooth to communicate between the earpiece and the cell phone. A Bluetooth-enabled GPS receiver receives the GPS satellite signals, calculates its position, then sends the position to a nearby display device. For example, a Bluetooth GPS receiver on the dash of a car or on the outside of the car can transmit position information to a map display device, such as a cell phone or a laptop computer, inside the car.

HP iPAQ Bluetooth GPS receiver and HP PDA.

GPS and the Laptop Computer

When a GPS receiver is used with a laptop computer, the receiver calculates and reports current position to the laptop, while the laptop runs the mapping software that displays position, routes, POIs and other navigational information. Connecting your receiver to your lap-

Garmin Bluetooth receiver.

External receiver for laptop attaches to roof of car.

top is not new, but what is new is the types of receivers developed specifically for use with laptop computers. Entire GPS receivers are manufactured in an external antenna. The combined GPS receiver/antenna is placed on the roof of the car and the wire from the receiver connects to the PC. These combination units are very small and convenient to use. No longer do you have to connect your much larger hand-held receiver to your laptop.

GPS receivers for laptops can also use Bluetooth to communicate with the laptop, or the GPS receiver may also come in PCMCIA form factor that plugs directly into the laptop; however, the PCMCIA receiver may not be able to pick up the satellite signals while in a vehicle. Current external antenna/GPS receiver combo models are the Garmin 18 USB and the Navman E Series, for laptops.

GPS for PDAs and Pocket PCs

As technology progresses, PDAs and Pocket PCs have become increasingly advanced and more powerful. Some of these devices are capable of running programs similar to a laptop and have operating systems of equal power. Using the Bluetooth technology discussed earlier, any of these systems can be turned into a powerful GPS unit with all the capabilities of road navigation units and great flexibility for loading electronic maps. Some PDAs and Pocket PCs come with built-in GPS receivers or can be attached to external docking units with GPS capability (e.g., www.pharosgps.com, TomTom Navigator, some HP iPAQ products, Garmin iQue, Navman i-Series).

External receiver with Bluetooth-enabled PDA.

Turning a PDA/Pocket PC into a GPS receiver requires:

• Specialized GPS receiver.
No display, no waypoint memory.

A Garmin iQue GPS-enabled PDA.

Connects to PDA/PocketPC by wire,
Bluetooth or Compact Flash (CF) insert.
Simply reports current position to the Pocket PC.
- Mapping software for PDA/Pocket PC.
The PDA/PocketPC runs the mapping software.
Shows current position on a map.
Remembers routes and waypoints.
Provides automatically generated maps.

PDA or Pocket PC navigation systems have advantages and disadvantages over traditional hand-held GPS receivers.

Garmin Compact Flash (CF) GPS receiver for inserting into PocketPC.

Advantages
- You already own the PDA/PocketPC and likely carry it everywhere.
- Color screen.
- Larger than average screen.
- More memory to store waypoints, routes, track logs and maps.
- Greater variety of map programs available.
- Ability to automatically calculate route using road software.

Disadvantages
- Uses lots of batteries. Best used in a vehicle with cigarette-lighter power.
- Not waterproof.
- More fragile.
- Larger than the smallest available hand-held receivers.
- Higher cost (if you do not already own a PDA/Pocket PC for other purposes).

Cellular Phones

Utilizing small GPS chips, many cellular phones are becoming GPS receivers (e.g., Pharos GPS phone, Sprint Nextel, Verizon). Other phones support map software and receive position information from a Bluetooth receiver (e.g., www.amazegps.com). Other cell phones include a GPS receiver to report position to mapping software hosted by the cellular provider. Using this type of service, the cellular phone transmits its GPS location data through the cellular network and receives route generation, voice prompts, turn-by-turn directions and POI information from a central location (e.g., Telenav). Garmin also provides GPS service for enabled phones. Features provided include the turn-by-turn routing, real-time traffic data, the weather forecast, and gas prices and nearby gas stations.

Tracking Devices

Tracking devices have become more available for the standard consumer and have many different applications. Using cellular GSM (Global System for Mobile communications, which is a cell phone standard) and SMS (i.e., Short Message Service, also known as text messaging) technology, small receivers can be used to keep track of employees (e.g., Nextel/Sprint AirClic® MP), children, the elderly (e.g., www.findware.co.uk), vehicles, even pets (Garmin Astro dog tracker). Most of these devices require a subscription service, but they allow you to create different notification settings, such as if the person leaves an area or travels faster than a certain speed. Notifications can be sent to your cellular phone in the form of SMS messages. You can also use a computer to locate the person being tracked and keep a record of their movements. Many of these receivers sense movement and slow down transmission of position information when the receiver is not moving, then increase the transmission rate when motion resumes.

GPS-equipped Two-way Radios

Consumer radios operate on the Family Radio Service (FRS) and/or General Mobile Radio Service (GMRS) bands. GPS receivers such as the Garmin Rhino series, combine a fully functioning GPS receiver with a radio that uses both the FRS and GMRS bands. In addition to allowing you to talk with the other members of your group, the Rhino displays on the Map screen the position of the other Rhino users in your group. The range is anywhere from 2 to 14 miles, depending on the power of the radio transmitter, surrounding terrain and atmospheric conditions.

Garmin Rhino GPS receiver/radio.

Astro GPS receiver and dog position transmitter.

Another application of radio equipped GPS receivers is geared toward hunters who hunt with dogs. The Garmin Astro, mentioned above, combines a small GPS receiver with a radio transmitter which can be attached to a dog's harness or collar. The dog's position is transmitted to the hunter's radio-equipped GPS receiver, allowing the hunter to track the dog through tall brush and foliage.

13 Other Grids

There are innumerable grids in addition to UTM and latitude/longitude. Even though most maps have latitude/longitude, it may be the secondary grid and the primary grid may be more convenient to use. Some grids are limited to a single country or area like the Universal Polar Stereographic or the Ordnance Survey Great Britain grid. If you think you will need a grid that is not UTM or latitude/longitude, be sure the receiver supports it before you buy. The chances are that a mid- to upper-price-range receiver will support any grid you will ever need, because receivers in those price ranges generally support over a hundred grids. This chapter describes the basics of the following grids:

- Universal Polar Stereographic (UPS)
- Ordnance Survey Great Britain (OSGB)
- Military Grid Reference System (MGRS)
- Maidenhead

Universal Polar Stereographic (UPS)

The Universal Polar Stereographic grid was developed to provide the Arctic and Antarctic regions with a uniform grid. The UPS grid is very much like the UTM grid. In fact, the UTM grid could be extended to cover the entire earth, though the coordinates would be confusing near the poles because the zones would be very narrow. Like UTM, the UPS grid has eastings and northings that form 1 km (0.62 mi.) squares. The UPS coordinates for both the North and South poles are given in the tables. Most receivers display the zone as 0. The Greenwich Meridian (0° longitude) and the International Date Line (W 180° longitude) form the zone meridian that references all easting measurements. The longitude lines W 90° and E 90° form the meridian for measuring northing coordinates. Some receivers use the MGRS letters Y and Z for the North Pole and A and B for the South Pole to label west and east of the easting meridian respectively. The numbers by the small circles in the figure correspond to the numbered coordinates listed below. There is a direct correspondence between the latitude/longitude grid and the UPS grid only at the meridians, but the tables include latitude/longitude for all the points to help you relate the UPS coordinate to a grid you already know.

Arctic UPS Coordinates

Point	Lat/Long	UPS Coordinate
#1	N 84°, W 180°	0 Y 2000000m.E. 2666760m.N.
#2	N 88°, W 135°	0 Y 1842965m.E. 2157035m.N.
#3	N 84°, W 90°	0 Y 1333237m.E. 2000000m.N.
#4	N 88°, W 45°	0 Y 1842965m.E. 1842965m.N.
#5	N 84°, E 0°	0 Z 2000000m.E. 1333237m.N.
#6	N 88°, E 45°	0 Z 2157035m.E. 1842965m.N.
#7	N 84°, E 90°	0 Z 2666764m.E. 2000000m.N.
#8	N 88°, E 135°	0 Z 2157035m.E. 2157035m.N.
North Pole N 90°		0 Z 2000000m.E. 2000000m.N.

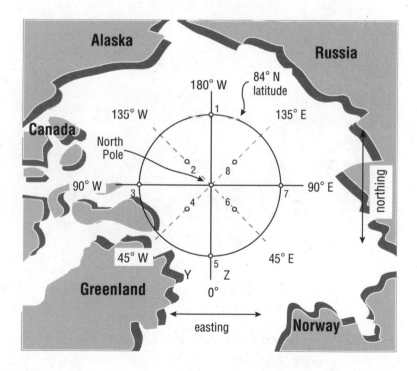

The UPS grid for the Antarctic region is similar to the UPS grid for the Arctic except the outer circle is latitude S 80°. The figure shows how the UPS grid is laid out for the South Pole and also gives the coordinates of some points indicated by numbers next to small circles in the figure.

Point	Lat/Long	UPS Coordinate	
#1	S 80°, W 180°	0A 2000000m.E.	0886989m.N.
#2	S 85°, W 135°	0A 1607211m.E.	1657211m.N.
#3	S 80°, W 90°	0A 0886989m.E.	2000000m.N.
#4	S 85°, W 45°	0A 1607211m.E.	2392789m.N.
#5	S 80°, E 0°	0B 2000000m.E.	3113011m.N.
#6	S 85°, E 45°	0B 2392789m.E.	2392789m.N.
#7	S 80°, E 90°	0B 3113011m.E.	2000000m.N.
#8	S 85°, E 135°	0B 2392789m.E.	1607311m.N.
South Pole S 90°		0B 2000000m.E.	2000000m.N.

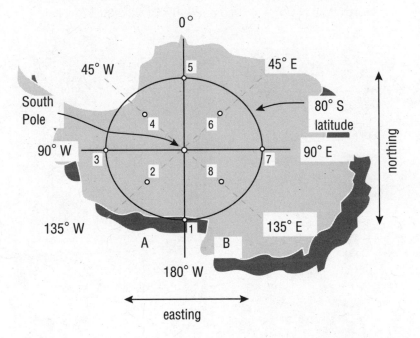

Ordnance Survey Great Britain (OSGB)

The Ordnance Survey Great Britain is a national grid that covers only the U.K. The country is divided into 100 km (62.1 mi.) square sections (100 km × 100 km) that are designated with two letters as shown in the figure. Each section is divided into 1 km squares. The grid uses easting and northing numbers to describe locations, like the UTM grid. A large-scale map (1:25,000 or 1:50,000) covers only a portion of a 100 km section.

Eastings and Northings

- OSGB easting and northing coordinates are much like UTM easting and northing coordinates.

- Increasing easting numbers means you are going east.

- Increasing northing numbers means you are going north.

- A full coordinate is **440**000m, **457**000m.

- The large numbers shown on a map are an abbreviation.
For example, 40 means **440**000m and 57 means **457**000m.

				HP			
HQ	HR	HS	HT	HU	JQ		
HV	HW	HX	HY	HZ	JV		
NA	NB	NC	ND	NE	OA		
NF	NG	NH	NJ	NK	OF		
NL	NM	NN	NO	NP	OL		
NQ	NR	NS	NT	NU	OQ		
	NW	NX	NY	NZ	OV		
	SB	SC	SD	SE	TA		
	SG	SH	SJ	SK	TL	TG	
	SM	SN	SO	SP	TL	TM	
SQ	SR	SS	ST	SU	TQ	TR	
SV	SW	SX	SY	SZ	TV		

- The distance between 40 and 41 is 1000 m (1 km).
- The last three numbers stand for meters.
- The distance between **339**000m and **339**541m is 541 m (1775 ft.).

OSGB Coordinates

- An OSGB coordinate includes section letters, easting and northing.

- The section is printed on the map. A typical value is SE.

- When section is specified, omit the small number in front of the coordinate.

- An OSGB coordinate is written as SE **38**000m, **57**900m.

- The coordinate may be abbreviated as SE **38**0 **57**9.

- Like UTM, the large numbers on the map are used to abbreviate.

- Eastings and northings are not marked m.E. and m.N.

- When using a GPS receiver, the full coordinate – section letters, easting and northing – must be used. Abbreviations are too map-specific.

If you travel in Great Britain and use your GPS receiver, you will find the OSGB grid is much easier to use than the latitude/longitude grid, because the 1 km (0.62 mi.) squares are printed on the map. On large scale maps, it is possible to read coordinates directly from the map. However, if you prefer to use a ruler, you can use any of those introduced in the UTM grid chapters if you have the correct scale.

Military Grid Reference System (MGRS)

Most outdoor enthusiasts do not use the MGRS grid, because USGS topographical maps do not provide it. However, if you prefer the MGRS grid, electronic map databases will print maps with it. MGRS is simply a modified form of the UTM grid in which the first two numbers of the easting and northing are replaced with letters. It is a lot like the OSGB grid because the letters are assigned to 100 km × 100 km squares.

The UTM and MGRS coordinates for the same location are shown below. The first two numbers of the easting and northing were converted to the letters "WB." The 05 from the easting became the "W," while the 36 from the northing became "B." The m.E. and m.N. were removed. All other numbers remain the same. Sometimes MGRS coordinates are written as a continuous string of numbers and letters as shown on the last line, but most GPS receivers keep the zone designator, the easting and the northing separate, so that everything is legible. Much like any other grid, GPS receivers require you to enter all the numbers of each coordinate, which means a position is specified down to 1 m.

UTM: 12 S   0501788m.E. 3690619m.N.

MGRS: 12 S   WB 01788, 90619

12 SWB 0178890619

There still are tables available that can be used to convert from UTM to MGRS by hand. But if you ever have a full UTM coordinate and need to convert it to MGRS, simply enter the UTM coordinate into the receiver, then switch the receiver to the MGRS grid. The receiver will do the conversion for you.

Maidenhead Grid

The Maidenhead grid was developed and is used by amateur radio operators. It divides the world into grids with dimensions 20° of longitude by 10° of latitude, which are identified by pairs of letters, AA–RR. The grids are subdivided into areas 2° x 1° and labeled with pairs of numbers, 00–99. The areas are further subdivided into subareas that are 5' of longitude by 2.5' of latitude and labeled with letter pairs AA–XX. A Maidenhead coordinate looks like this: EM18BX.

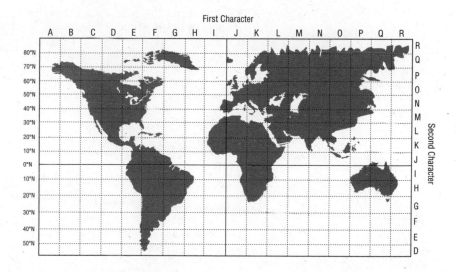

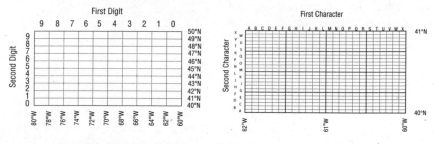

Maidenhead area subdivisions.   *Maidenhead subareas.*

Glossary

2D Mode. Position calculations in two dimensions. In terms of a GPS receiver, it means the receiver can lock onto only three satellites, so it cannot provide altitude. There may be substantial error in the horizontal coordinate it does provide.

3D Mode. Position calculations in three dimensions. The GPS receiver has locked on to four satellites. It provides a vertical as well as a horizontal coordinate.

Almanac Data. Satellite position information. Each satellite broadcasts the position information for all the satellites. The receiver stores this information so it can determine its own position. It takes about 12.5 minutes for the satellites to transfer the position data to the receiver.

Altimeter. A device that measures distance above sea level. Atmospheric pressure decreases as you rise in altitude, so most altimeters measure atmospheric pressure and relate it to height above sea level.

Antenna. A receiver needs an antenna to pick up the satellite signals beamed down from space. There are two common types for hand-held receivers: patch (microstrip) and quadrifilar helix. The antenna is one of the most important components of a receiver. A remote antenna is separate from the antenna built into the receiver and is usually connected to the receiver with a cable. An active remote antenna is one that amplifies the satellite signals before sending them through the cable to the receiver.

Anti-spoofing. It is possible to confuse GPS receivers by transmitting signals that look similar to the real satellite signals. Such an attack is known as spoofing. The military countermeasure is to encrypt the P code so only authorized users can recognize it and can detect and reject faked signals. See also **Spoofing**.

Azimuth. The direction of travel or the direction between two points in reference to true north or magnetic north. When expressed in degrees, its value ranges from 0° to 360°. A compass heading is an azimuth. In most places, the word "bearing" has grown to mean the same thing as azimuth. See **Bearing**.

Bearing. A bearing is your direction of travel or the direction between two points. Like an azimuth, a bearing is measured in reference to true north or magnetic north, but its value never goes over 90°. A bearing is always measured from the cardinal directions north or south. A typical bearing would be N45°E, which is the same as an azimuth of 45°. The bearing S45°W is an azimuth of 225°. The use of the word "bearing" has changed over the years and now means the same thing as "azimuth."

Bluetooth. Bluetooth enables electronic devices to communicate without wires. It is a short-range radio system. The communicating electronic devices must both have Bluetooth circuits to transmit and receive signals. The main advantage is that there is no wire to get in the way. Bluetooth devices have a range of about 10 m (32.8 ft.).

Channel. The part of the GPS receiver's electronics that tunes in on a satellite's signal and sends the resulting information to the receiver's processor for position calculation.

Chart. A map of waterways or airways.

Coarse Acquisition Code. The GPS satellites send two distinct signals: precision code (P code) and coarse acquisition code (CA code). Civilian receivers use CA code to determine position. Military receivers use CA code to synchronize to the P code before switching to P code exclusively. Selective Availability used to affect the CA code and thereby the accuracy of civilian receivers. The CA code is transmitted on only one radio frequency, so it is impossible for a civilian receiver to detect the delay through the ionosphere. The accuracy provided by CA code is called the Standard Positioning Service (SPS).

Codeless Receivers. A class of GPS receivers that do not use P code or CA code to determine position. Codeless receivers measure the change in modulation in the satellite radio waves. They use sophisticated signal processing techniques to make position measurements accurate to centimeters. It can take days to make a single measurement.

Cold Start. A receiver experiences a cold start when it has to download the almanac information from the satellites before it can begin to calculate its own position. See **Time to First Fix**.

Coordinate. The numbers and letters that describe a position. Every position on earth has a unique coordinate. The coordinate system determines the grid and how the coordinate is written.

Course. The path between two points. GPS receivers always indicate the straight line between two points.

Course Deviation Indicator (CDI). A method for displaying the amount and direction of CrossTrack Error (XTE).

Course Made Good (CMG). The bearing from your starting point to your present position.

Course Over Ground (COG). Same as Course Made Good (CMG).

CrossTrack Error (XTE). The distance between your present position and the straight-line course between two points. It is the amount you are off the desired track (DTK).

Declination. The difference, in degrees or mils, between the north pole and the magnetic pole from your position. Many receivers have tables in their memory that tell them the amount and direction of declination for any position on earth, which means the receiver, once locked on to the satellites, can automatically convert true north bearings to magnetic bearings and vice versa.

Degree. A part of a circle. The degree divides the circle into 360 even pieces. Bearings are also expressed in degrees. Degrees are subdivided into 60 minutes, which in turn are split into 60 1-second intervals.

DGPS Ready. A receiver is DGPS ready if it is capable of accepting Differential GPS correction data and using it to make its own position calculation more accurate. Additional equipment must be connected to the receiver to pick up the correction radio transmissions.

Differential GPS (DGPS). A method of improving civilian receiver accuracy. DGPS can be accurate to 15 m and below. DGPS corrections can be made instantaneously (real-time) as you are traveling or after the trip on stored waypoints (post-processing).

Dilution of Precision (DOP). An analysis of the satellite geometry and its impact on accuracy. Some satellite geometries provide more accurate position calculations. The receiver measures several factors that dilute the position accuracy, and adds them all together to estimate how much error is present in its position calculation. Components of the DOP are horizontal, vertical, position and time dilutions of precision. A low value for a DOP means the receiver can accurately make a position calculation. A high value means there is increasingly more error in the position reported. Good DOP values range between 1 and 3. Most receivers will not even try to calculate position if the DOP values are greater than 6.

Easting. The distance east or west from the zone meridian. Easting coordinates are used in several grid systems. UTM, OSGB and MGRS are a few.

Ephemeris. The path and orbit information for a specific satellite. Selective Availability used to truncate the ephemeris information to limit the civilian receiver's accuracy.

Estimated Position Error (EPE). Many receivers report the potential error of a position calculation. The receiver knows the satellite geometry, and using the DOP values it estimates the amount of error that may be present in the position it calculates.

Estimated Time En Route (ETE). The amount of time remaining to arrive at the destination. ETE depends on the speed you are going directly toward the destination, which is called Velocity Made Good (VMG). If you are traveling away from the destination, the ETE cannot be calculated, because you will never arrive.

Estimated Time of Arrival (ETA). The time of the day of arrival at the destination. ETA depends on the speed you are going directly toward the destination, which is called Velocity Made Good (VMG). If you are traveling away from the destination, the ETA cannot be calculated, because you will never arrive.

Global Positioning System (GPS). A system of 24 satellites that allows a GPS receiver to determine its position anywhere in the world. There are two types of receivers: military and civilian. Military receivers are always accurate to about 1 m (3.3 ft.). Civilian receivers were made less accurate by Selective Availability, but now that that has been removed, they are accurate to at least 15 m (49.2 ft.).

GLONASS. The Russian equivalent of the U.S. GPS. Its full name is Global'naya Navigatsionnaya Sputnikovaya Sistema.

Goto Function. A mode of operation where the receiver guides you to a destination. You must have previously stored the destination's coordinate in the receiver's memory. The receiver uses the satellite signals to find its present position, then calculates the bearing and distance to the destination. In Goto mode, the receiver usually displays a screen that points the direction you need to travel to arrive at the destination.

Greenwich Mean Time (GMT). The time as measured from Greenwich, England, or 0° longitude. Refer to **Universal Time Coordinated**.

Grid. The horizontal and vertical lines on a map that fix your position. There are a lot of different grid systems because there are many different ways of translating a position from a sphere to a flat map. The most common grid systems are Universal Transverse Mercator (UTM) and latitude/longitude.

Grid North. The orientation of a map's grid. Cartographers try to align the vertical lines on the map with true north. However, there is usually a small difference between grid north and true north across the map, but the difference is so slight that it can usually be ignored for land navigation.

Ground Speed. Your speed across the ground regardless of direction. The speedometer in a car measures ground speed.

Heading. The direction of travel expressed as either a magnetic north or a true north bearing.

Horizontal Dilution of Precision (HDOP). See **Dilution of Precision**.

Ionosphere. A layer of the earth's atmosphere between 50 and 250 mi. above the surface. The GPS satellite signals are delayed as they pass through the ionosphere. If the effect of the delay is not removed or compensated for, the receiver's position calculation is inaccurate.

L1 and L2. The P code is transmitted on two radio frequencies known as L1 and L2: 1575.42 MHz and 1227.6 MHz respectively. The CA code is transmitted only on L1.

Landmark. Same as a **waypoint**. It can also refer to a distinct landform that is easily recognizable.

Latitude/Longitude. A spherical coordinate system. The lines of latitude and longitude form a grid system used to fix position. Latitude lines run parallel to the equator and measure distance from the equator, while longitude lines are drawn from pole to pole and measure distance from the prime meridian at Greenwich, England (0°). Coordinates are measured in degrees, minutes or seconds.

Local Area Augmentation System (LAAS). A real-time DGPS correction system being developed by the U.S. Federal Aviation Administration for use around airports. It will provide greater GPS accuracy than WAAS.

Lock. A receiver is locked when it can detect three or more satellites and it can use their signals to determine its own position.

Magnetic Declination. See **Declination**.

Magnetic North. The direction the compass needle points. A compass needle always points toward the magnetic pole, located on Bathurst Island in northern Canada. The magnetic pole is not the north pole.

Man-overboard (MOB). A GPS receiver function that allows you to quickly mark a position in an emergency

Map Datum. All maps are drawn with respect to a reference point. The reference point is called the datum. Most map datum only cover a portion of the earth, like the North American Datum

of 1927 (NAD 27), which covers only the continent of North America. The GPS makes it possible to have a worldwide datum like the World Geodetic System of 1984 (WGS 84).

Map Scales. The scale of a map is usually expressed in the form 1:24,000. The scale means that every inch on the map represents 24,000 inches on the ground. A large-scale map is one that is zoomed in, whereas a small-scale map covers a lot of area on a single page.

Meridian. A longitude line that is used as a reference. The longitude line through Greenwich, England, is referred to as the prime meridian and is labeled 0°. All other longitude lines are measured in relation to the prime meridian. Each zone in the UTM system also has a zone meridian used as the reference point for all east/west measurements.

Mil. A part of a circle. When a circle is divided into 6400 equal-sized pieces, a mil is one of those pieces.

Military Grid Reference System (MGRS). The grid system used by the U.S. military. It is similar to UTM except it replaces the most significant digits of the easting and northing numbers with letters. Some mapping programs will put the MGRS grid on USGS topographical maps for use by civilians.

Multipath. When the same signal from a satellite enters a receiver's antenna from more than one direction, it is called multipath. Usually the radio waves travel straight from the satellite to the receiver, but if they happen to bounce off some hard object, they will enter as both a direct signal and a reflected signal.

National Maritime Electronics Association (NMEA). NMEA protocols (how data is sent and what format it uses) specify the type and order of data sent and received by navigation equipment. If two pieces of equipment use the same NMEA protocol, they will understand each other and operate together.

Navstar. Navstar Global Positioning System was the original name for the navigation system, but the Navstar part was soon lost and the system is known today only as GPS. See **Global Positioning System**.

Northing. The distance north or south of a fixed reference point. The UTM system uses the equator as the reference. Northing coordinates are used in several grid systems. UTM, OSGB and MGRS are a few.

Off Course. A navigational statistic that tells you how far you are from the straight-line course between two points. It is usually reported as a distance to the right or left of the straight-line course.

Outage. An outage occurs when the satellite geometry is so poor that the receiver cannot make an accurate position calculation. Most receivers will not lock when the position dilution of precision is greater than six. See also **Dilution of Precision**.

Point-to-Point Calculation. The calculation of bearing and distance between two points.

Position Dilution of Precision (PDOP). See **Dilution of Precision**.

Precision Code. The GPS satellites send two distinct signals: precision code (P code) and coarse acquisition code (CA code). Civilian receivers use CA code to determine position. Military receivers use CA code to synchronize to the P code before switching to P code exclusively. During the time Selective Availability was used, it did not affect the P code. P code is transmitted from space on two different frequencies, which enable military receivers to detect and eliminate propagation delays introduced in the ionosphere. The accuracy provided by the P code is called the Precise Positioning Service (PPS).

Ranging. Ranging is the technique used in the GPS for a receiver to measure its distance to a satellite.

Route Function. A list of sequential waypoints. The GPS receiver guides you from the first waypoint on the list to each point in order until you arrive at the destination. See also **Goto Function**.

RS-232. A standard type of connection to a computer. It is a serial port that allows communication between a computer and a receiver. A special cable connects the computer's RS-232 port to the receiver. See also **Universal Serial Bus (USB)**

Satellite Geometry. The positions of the satellites in the sky relative to your position on earth. The best satellite geometry is one satellite overhead with the others spread evenly around the horizon. See also **Dilution of Precision**.

Selective Availability (SA). The technique formerly used by the U.S. Department of Defense to make civilian receivers less accurate. It limited horizontal accuracy to between 15 and 100 m (49.2 and 328 ft.) and vertical accuracy to 156 m (512 ft.). Selective Availability was eliminated May 2, 2000.

Speed of Advance (SOA). Same as Velocity Made Good.

Speed Over Ground (SOG). The speed you are traveling regardless of direction. It is the same as ground speed.

Spoofing. Spoofing is a method of attacking the GPS to render it useless. The attacker transmits radio signals at the same frequency as the GPS signals so that the receiver mistakes the fake signal for the real one and calculates an incorrect position. The countermeasure to spoofing is called anti-spoofing. Spoofing can be detected and foiled only by military receivers, not civilian ones.

Time to First Fix (TTFF). The amount of time (about 12 minutes) it takes a receiver to make its first position fix after it has been off for several months, has lost memory or has been moved more than 480 km (300 mi). Before the receiver can calculate its position, it needs to download all the position information about every satellite.

True North. The direction to the North Pole. The North Pole is not the magnetic pole. The difference in direction between the North Pole and the magnetic pole is called declination.

Universal Polar Stereographic Grid (UPS). The grid that covers the Arctic and Antarctic regions. It is similar to UTM, with eastings and northings.

Universal Serial Bus (USB). An electronic connection associated with computers. The USB allows a computer user to connect a GPS receiver or memory card programmer to a computer to transfer data. USB offers faster data transfer than a computer's serial bus; so if you have a choice between a USB and an RS-232 serial connection, select USB.

Universal Time Coordinated (UTC). Essentially Greenwich Mean Time. GPS time, as maintained by the satellites, is converted to UTC inside the receiver.

Universal Transverse Mercator Grid (UTM). The grid that splits the earth into 60 zones, each 6° wide. Its coordinates are relative to the equator and a zone meridian and are called eastings and northings. The UTM grid is used only between North 84° and South 80° because the UPS grid already provides a uniform grid for the poles.

Velocity Made Good (VMG). Your speed toward the destination. If you are traveling directly toward the destination, VMG is the same as your ground speed. If you are not on course, the VMG is less than your ground speed. If you are headed away from the destination, VMG is zero regardless of how fast you are going.

Waterproof. A receiver is listed as waterproof if it can be completely submerged in water without being ruined.

Water Resistant. A water-resistant receiver can be used in a damp environment, but it is not designed to be submerged or get really wet.

Waypoint. The coordinates of a location. Waypoints are stored in the receiver's memory. You can store your present position, as determined by the receiver, as a waypoint or you can store the position of any place in the world by reading its coordinate from a map and typing it into the receiver.

Wide Area Augmentation System (WAAS). Real-time DGPS corrections transmitted by satellites for use by receivers in the continental U.S., Alaska and parts of southern Canada. WAAS increases GPS receiver accuracy from 15 m (49.2 ft) to 3 m (9.8 ft.).

Y Code. The encrypted version of P code. See **Anti-spoofing**.

Other Books

Some basic map and compass books:

Be Expert with Map and Compass: The Complete Orienteering Handbook, Bjorn Kjellstrom, Macmillan General Reference, 1994

Staying Found: The Complete Map and Compass Handbook, 3rd ed., June Fleming, Mountaineers Books, 2001

The Basic Essentials of Map and Compass, 3rd ed., Cliff Jacobson, FalconGuide, 2007

The Outward Bound Map and Compass Handbook, rev. ed., Glenn Randall, Lyons Press, 1998

An excellent sourcebook of maps for all over the world:

The Map Catalog: Every Kind of Map and Chart on Earth and Even Some Above It, 3rd ed., Joel Makower, ed., Vintage Books, 1992

Some heavy-duty books about mapmaking and GPS:

Global Positioning System: Theory and Practice, 5th, rev. ed., B. Hofmann-Wellenhof et al., Springer-Verlag, 2001

Map Use: Reading, Analysis, and Interpretation, 5th ed., Phillip C. Muehrcke, JP Publications, 2005

The Navstar Global Positioning System, Tom Logsdon, Springer, 2007

Understanding the Navstar: GPS, GIS and IVHS, 2nd ed., Tom Logsdon, Springer, 1995

A fine book about Amerigo Vespucci; short but interesting:

Forgotten Voyager: The Story of Amerigo Vespucci, Ann Fitzpatrick Alper, Carolrhoda Books, 1991.

At the time of publication, the following Internet address was a good source of GPS information:

www.colorado.edu/geography/gcraft/notes/gps/gps_f.html

Index

Lawrence Letham, a registered patent attorney, electrical engineer and outdoor enthusiast, has long been fascinated by maps and the art of navigation. Alex Letham, a bioengineer and now first-year medical student, mountain bikes, hikes, kayaks and snowboards regularly. Both live in Arizona and love their GPS receivers.